38p

Alimentary system	Eyes	Ears	Flexure				Tail bud and limb buds
oregut 0.15 mm. long.							
oregut 0.3 to 0.4 mm. long.							
oregut 0.5 to 0.8 mm. long.	Primary **optic** vesicles form.						
oregut about 1.0 mm. long.	Optic stalks begin to constrict at the base of the primary vesicles.		The head bends ventrally and sinks into the yolk.				
oregut is about 1.3 mm. long.	Differentiated into optic stalk and optic vesicle.	The auditory placodes begin to form from thickened ectoderm.	This cranial flexure increases in the region of the mid-brain.	The first signs of torsion appear in the head region.	The head amniotic fold begins to rise up and grow back.		
he foregut is 1.5 mm. long nd there are indications of e first pair of visceral clefts.	The ectoderm outside the primary vesicles thickens and becomes the rudiment of the lens.	The auditory placodes invaginate to form auditory pits.	Further sinking of the head into the yolk is prevented by the head twisting (torsion). Cranial flexure approaches 90°.	The head turns on to the left side. This torsion may reach the first two or three somites.	The head amniotic fold has grown back over the fore-brain.		The remains of the primitive streak begin contributing material posteriorly to the tail bud.
e first pair of visceral clefts e distinct and the second ir begin to form.	The lens rudiments invaginate to form lens vesicles. The optic vesicles invaginate to form optic cups	The mouth of each auditory pit begins to constrict and auditory vesicles form.	Cranial flexure, i.e. angle between fore- and hind-brain, is 90°. Cervical flexure begins in hind-brain region and trunk flexure can also be seen.	The head is fully turned to the left. The first five to seven somites also exhibit torsion.	The hind brain and first few somites are covered by the amnion. Tail folds may begin to develop.		There is a distinct tail bud.
e first and second visceral fts are clearly visible; the rd pair begin to develop. e hind gut appears.	The mouth of the lens vesicle begins to close.	The mouth of the auditory vesicle is reduced to a small aperture.	Cranial flexure causes the fore-brain to be directed backwards close to the heart. Cervical flexure becomes a broad curve.	Torsion is apparent in somites eight to ten.	The sero-amniotic connection is somewhat attenuated. The amnion covers somites six to thirteen. Tail fold appears.		The tail bud begins to develop posterior to the hind gut.
e first, second and third ceral clefts are present. e liver bud appears as do e tail gut and anal plate.	The lens becomes cut off from the ectoderm. The optic cups are almost closed. The retina distinct. The eye is still anterior to the ear.	The auditory vesicle is connected to the small ectodermal aperture by the ductus endolymphaticus.	Cranial flexure is at its maximum. Cervical flexure increases. Trunk flexure is noticeable in the region of somites ten to twelve.	Torsion extends to somites eleven, twelve and thirteen or even further.	The head fold grows back and may lie anywhere between somites ten and eighteen. The tail fold begins to grow forward.		The tail bud can be seen projecting behind the hind gut. The limb buds appear as low swellings.
e fourth pair of visceral fts develop. The liver bulge now conspicuous. Tail gut tends further into tail. baca begins to form.	The optic cup closes. The eyes now lie posterior to the ears.	The aperture closes.	Cranial flexure remains unchanged. Cervical flexure is about 100°. Trunk flexure develops into a broad curve. Caudal flexure begins.	Torsion as far back as somites fifteen to nineteen.	The head fold has extended to the region between somites sixteen to twenty-four. The tail fold has grown forward over somites 29–30.	Begins as an outgrowth of the hind gut in the cloacal region.	The tail bud begins to curve forward. The fore limb bud lies between somites 17–19 and the hind limb bud between somites 26–30.
ur pairs of visceral clefts. e tail gut begins to degener-. The anterior and posterior estinal portals approach h other, leaving an open estinal umbilicus of 3 mm. ng buds develop.	The eyes, due to flexure, lie well posterior to the ears.	The auditory vesicle is pear-shaped with a narrow ductus endolymphaticus.	Caudal flexure causes tail to be at an angle of 90° to the body.	The whole posterior region exhibits some degree of torsion.	The head and tail folds meet or leave a small oval aperture over somites 26–28.	Allantoic stalk and vesicle. The vesicle enlarges after 72 hours.	Limb buds are now quite conspicuous and begin to exhibit nipple-shaped apices The hind limb bud extends to somite 32.

AN ATLAS OF
Embryology

By W. H. Freeman and Brian Bracegirdle

AN ATLAS OF EMBRYOLOGY
AN ATLAS OF HISTOLOGY
AN ADVANCED ATLAS OF HISTOLOGY
AN ATLAS OF INVERTEBRATE STRUCTURE

By Brian Bracegirdle and Patricia H. Miles

AN ATLAS OF PLANT STRUCTURE, Vol. 1
AN ATLAS OF PLANT STRUCTURE, Vol. 2
AN ATLAS OF CHORDATE STRUCTURE

AN ATLAS OF
Embryology

W H Freeman *BSc FIBiol*
latterly Head of Biology Department
Chislehurst and Sidcup Grammar School
latterly Chief Examiner 'A'-level Zoology, University of London

Brian Bracegirdle *BSc PhD FIBiol FRPS*

Third Edition

Heinemann Educational Books
London

Heinemann Educational Books Ltd.
22 Bedford Square, London WC1B 3HH

LONDON EDINBURGH MELBOURNE AUCKLAND
HONG KONG SINGAPORE KUALA LUMPUR NEW DELHI
IBADAN NAIROBI JOHANNESBURG
EXETER (NH) KINGSTON PORT OF SPAIN

ISBN 0 435 60318 3
© W. H. Freeman and Brian Bracegirdle
1963, 1967, 1978
First published 1963
Reprinted twice
Second Edition 1967
Reprinted four times
Third Edition 1978
Reprinted 1982

*The frontispiece shows
serial sections of 72 hour chick*

Printed in Great Britain by
Fletcher & Son Ltd, Norwich and bound by
Richard Clay (The Chaucer Press) Ltd, Bungay, Suffolk

Preface to the third edition

For this edition all the photographs have been remade, especially those of amphioxus for which an interference microscope was used to secure greater clarity. Additionally, we have included mammalian material.

   We would like to thank Professor T. W. Glenister for permitting us to publish photographs 84, 85, 87 and 89. These are photomicrographs of embryos in the collection of the Department of Anatomy of Charing Cross Hospital Medicine School, London.

W.H.F.
B.B.

February 1978

Preface to the first edition

This book consists of photomicrographs of sectioned and entire embryos of frog and chick, with large detailed drawings to correspond.

   Descriptive embryology is still recognised as a necessary and valuable part of courses in zoology and biology leading to the General Certificate of Education at Advanced Level, and to first degrees. As teachers and examiners we have become aware of the difficulties experienced by students in interpreting the embryological structures seen under the microscope. The present book is intended to help overcome these difficulties, while at the same time summarising the descriptive embryology of frog and chick in sufficient detail for degree level. Care has been taken to label fully, and to make the drawings and photographs large enough for clearness.

   It has become apparent that the embryology slides in general use are not of very high quality. For this reason, little attempt was made to obtain slides of better quality, but to use those normally confronting the student — in this way we hope to have improved the chances of artifacts being recognised as such. A large number of slides was looked through, but in the end we confined ourselves to a relatively small number of the more typical specimens. By doing this we were able to produce a book inexpensive enough for wide general use.

   Each slide was photographed through the microscope, with special attention being paid to securing a flat field and good depth of focus — especially difficult with these rather large specimens. Not all the slides selected for inclusion were of a quality desirable for photomicrography, as will be obvious from the photomicrographs themselves; but we feel that this need be no great drawback, since students are often required to interpret these poorer-quality slides.

Each drawing was made completely independently of the photograph, directly from the slide. An accurate outline was obtained by microprojection, with the emphasis on line work, as it should be in students' drawings. Where it made for greater clarity, the drawing was diagrammatised, as in the case of some of the embryonic membranes. Later the drawing was compared with the photograph, and dotting was added where it seemed desirable for greater clarity. It will be seen that more detail appears in many of the drawings than in the corresponding photographs. This detail is obtainable only by the proper use of the fine focusing of the microscope at increased magnification, and should serve as a salutary reminder to the student that it is necessary for *him* to do the same to interpret *his* slides!

Much care and effort has been expended on the labelling of the drawings, and all the usual texts have been consulted. Even so, it was often necessary to have recourse to serial sections, where these were available. In many cases, none were, and so some errors are likely to remain, even though we were fortunate to have the fullest co-operation of Dr Ruth Bellairs, of University College, London, in checking the work. We are most grateful to Dr Bellairs for her great help; any errors remaining are, of course, the entire responsibility of the authors.

It would not have been possible to have produced this work from the slides already in our possession. For their kindness in making available extra material, we are deeply indebted to the following: Mr Charles Biddolph, Mr C. V. Brewer, Dr Ben Dawes, Mrs J. Froud, Mr George Gardener, Mr A. T. Green, Mr C. Heather, Dr Brian Lofts, Mr C. T. Pugsley, Mr A. R. Tindall, Mr H. Whate, and the Zoology Department of Wye College. To Mr George Gardener we owe an additional debt for his early criticism and encouragement. We were likewise fortunate in our lettering artist, Mr Alan Plummer, who co-operated in a most wholehearted manner; and also in our Publishers — in Mr Alan Hill and Mr Hamish MacGibbon we found a most sympathetic support and facilitation of our aims. Last, but very definitely not least, we must thank our wives very sincerely indeed for their help and encouragement, and for their stoicism when surrounded for weeks on end by all the impedimenta of drawing and photomicrography.

W.H.F.
B.B.

September 1962

Colour Transparencies for Projection

Every photograph in this book has matching 2 × 2 colour transparencies for projection available from Philip Harris Biological Ltd, Oldmixon, Weston-super-Mare, Avon.

Each original master transparency was made at the same time as the negative used for the photograph in this book, exclusively for this Company. The authors recommend these slides for their quality and moderate cost as excellent aids to the teaching of embryology especially in conjunction with this book.

CONTENTS

46. Chick: blastoderm, 6-somite, E.
47. Chick: blastoderm, 10-somite, E.
48. Chick: blastoderm, 13-somite, E.
49. Chick: blastoderm, 17-somite, E.
50. Chick: blastoderm, 20-somite, E.
51. Chick: blastoderm, 25-somite, E.
52. Chick: blastoderm, 30-somite, E.
53. Chick: blastoderm, 35-somite, E.
54. Chick: blastoderm, 40-somite, E.
55. Chick: 6-somite stage, head region, T.S.
56. Chick: 6-somite stage, somitic region, T.S.
57. Chick: 6-somite stage, notochord, T.S.
58. Chick: 6-somite stage, primitive streak, T.S.
59. Chick: 6-somite stage, V.L.S.
60. Chick: 10-somite stage, V.L.S.
61. Chick: 10-somite stage, forebrain region, T.S.
62. Chick: 10-somite stage, hindbrain region, T.S.
63. Chick: 10-somite stage, heart region, T.S.
64. Chick: 13-somite stage, heart region, T.S.
65. Chick: 13-somite stage, posterior trunk region, T.S.
66. Chick: 17-somite stage, trunk region, T.S.
67. Chick: 21-somite stage, trunk region, T.S.
68. Chick: 24-somite stage, fore and hind brain, T.S.
69. Chick: 24-somite stage, fore and hind brain, T.S.
70. Chick: 27-somite stage, trunk region, T.S.
71. Chick: 27-somite stage, posterior trunk region, T.S.
72. Chick: 27-somite stage, eye and ear region, T.S.
73. Chick: 30-somite stage, fore and hind brain, T.S.
74. Chick: 30-somite stage, heart region, T.S.
75. Chick: 30-somite stage, anterior trunk region, T.S.
76. Chick: 30-somite stage, posterior trunk region, T.S.
77. Chick: 36-somite stage, pharyngeal region, T.S.
78. Chick: 36-somite stage, hind-brain region, T.S.
79. Chick: 36-somite stage, olfactory pit region, T.S.
80. Chick: 36-somite stage, optic region, T.S.
81. Chick: 36-somite stage, trunk region, T.S.
82. Chick: 45-somite stage, tail and hind-limb region, T.S.
83. Chick: 36-somite stage, H.L.S.
84. Man: fertilised ovum, V.S.
85. Cow: cleavage, four-cell stage, T.S.
86. Rabbit: cleavage, morula in oviduct, T.S.
87. Guinea pig: blastocyst in endometrium of uterus, V.S.
88. Man: previllous trophoblast implanted in endometrium of uterus, 9 to 10 days, V.S.
89. Man: formation of the amnion and yolk sac, 13 days, V.S.
90. Man: primitive streak and allantois, 18 days, T.S.
91. Guinea pig: placenta of the labyrinthine type, V.S.
92. Man: foetal heart, 50 days, V.S.
93. Man: 50-day foetus, W.M.
94. Man: foetus, alizarin preparation to demonstrate bone formation, W.M.

A reference table of chick development is printed on the endpapers

AN ATLAS OF
Embryology

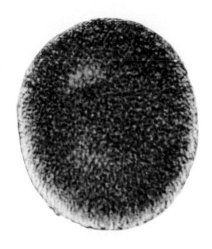

1. **Amphioxus:** cleavage,
1-cell stage, E. *mag. 450×*

2. **Amphioxus:** cleavage, 2-cell stage, E. *mag. 450×*

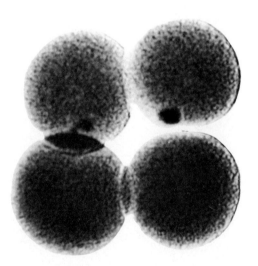

3. **Amphioxus:** cleavage, 4-cell
stage, E. *mag. 450×*

4. **Amphioxus:** cleavage,
8-cell stage, E. *mag. 450×*

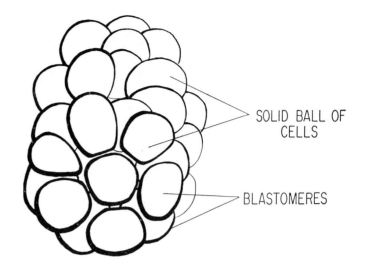

SOLID BALL OF
CELLS

BLASTOMERES

5. **Amphioxus:** cleavage,
morula, E. *mag. 450×*

5. Diagram of a Morula based on Specimen 5.

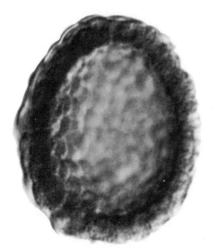

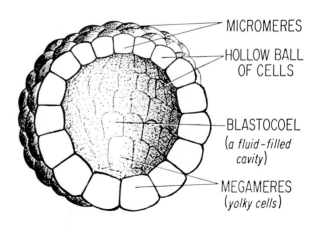

MICROMERES

HOLLOW BALL
OF CELLS

BLASTOCOEL
(*a fluid-filled
cavity*)

MEGAMERES
(*yolky cells*)

6. **Amphioxus:** cleavage,
blastula, E. *mag. 450×*

6. Diagram of Blastula, based on Specimen 6.

7. **Amphioxus:** cleavage,
blastula, V.S. *mag. 450×*

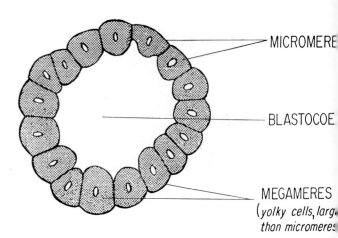

MICROMERE

BLASTOCOE

MEGAMERES
(*yolky cells, larg*
than micromere

7. Drawing of a V.S. of a blastula based on Specimen 7.

8. **Amphioxus:** gastrula,
E. *mag. 450×*

ECTODERM

DORSAL LIP OF
BLASTOPORE

BLASTOCOEL

INVAGINATING
ENDODERM

VENTRAL LIP

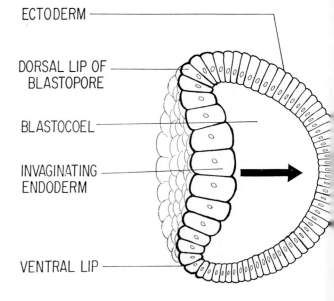

8. Diagram of Gastrula, based on Specimen 8.

9. **Amphioxus:** gastrula,
V.S. *mag. 450×*

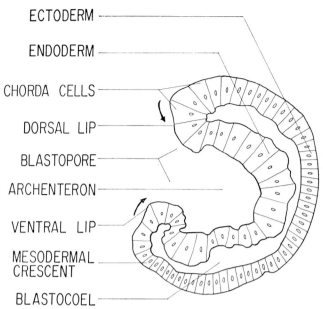

ECTODERM

ENDODERM

CHORDA CELLS

DORSAL LIP

BLASTOPORE

ARCHENTERON

VENTRAL LIP

MESODERMAL
CRESCENT

BLASTOCOEL

9. Diagram of Section through Gastrula, based on
Specimen 9.

10. **Amphioxus:** early embryo, E.
mag. 450×

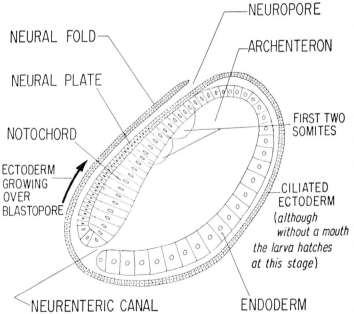

NEUROPORE

NEURAL FOLD

ARCHENTERON

NEURAL PLATE

FIRST TWO
SOMITES

NOTOCHORD

ECTODERM
GROWING
OVER
BLASTOPORE

CILIATED
ECTODERM
(*although
without a mouth
the larva hatches
at this stage*)

NEURENTERIC CANAL

ENDODERM

10. Diagram of early embryo based on Specimen 10.

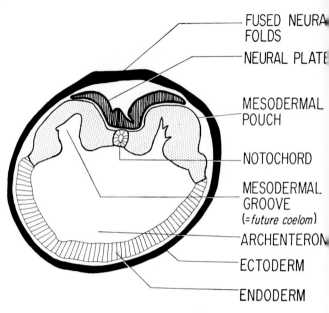

	FUSED NEURAL FOLDS
	NEURAL PLATE
	MESODERMAL POUCH
	NOTOCHORD
	MESODERMAL GROOVE (=*future coelom*)
	ARCHENTERON
	ECTODERM
	ENDODERM

11. **Amphioxus:** 8-somite stage, T.S. *mag. 450×*

11. Diagram of a T.S. of a 8-somite stage based on Specimen 11.

12. **Amphioxus:** 12-somite stage, E. *mag. 420×*

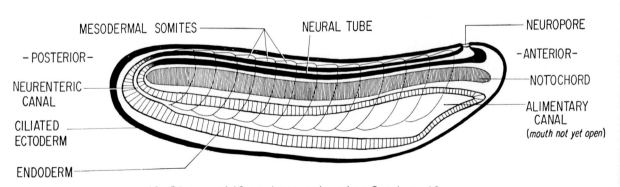

MESODERMAL SOMITES — NEURAL TUBE — NEUROPORE

– POSTERIOR – — ANTERIOR –

NEURENTERIC CANAL — NOTOCHORD

CILIATED ECTODERM — ALIMENTARY CANAL (*mouth not yet open*)

ENDODERM

12. Diagram of 12-somite stage based on Specimen 12.

13. **Amphioxus:** 17-somite stage, E. *mag. 420×*

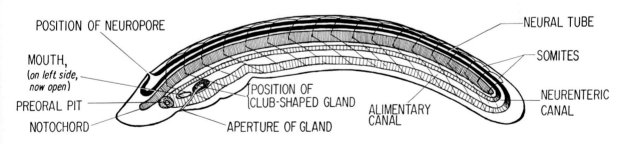

13. Diagram of a 17-somite stage based on Specimen 13.

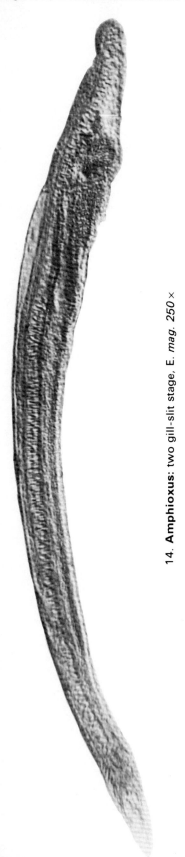

14. **Amphioxus**: two gill-slit stage. E. *mag. 250* ×

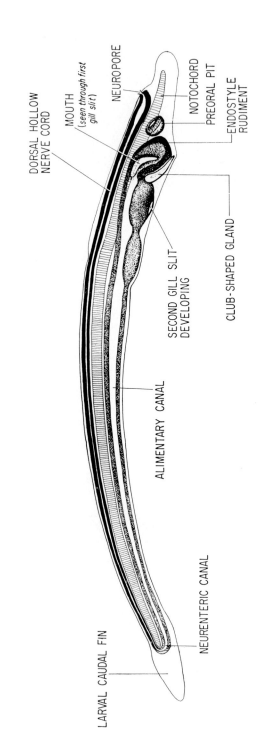

14. Diagram of two gill-slit stage based on Specimen 14.

9

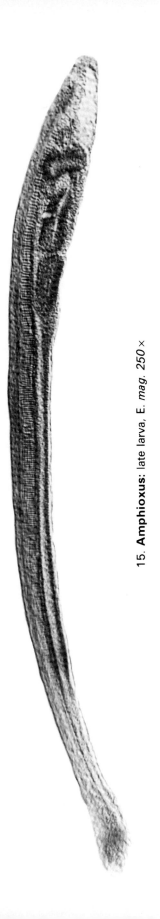

15. **Amphioxus:** late larva, E. *mag. 250* ×

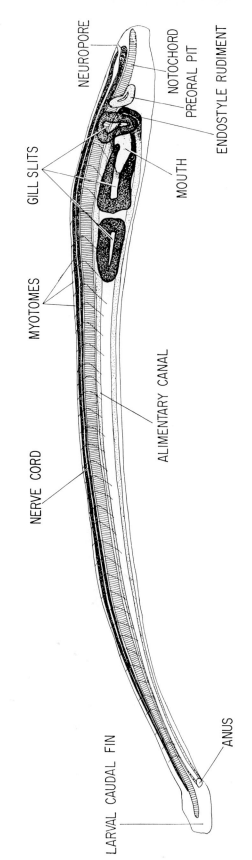

NEUROPORE

NOTOCHORD

PREORAL PIT

ENDOSTYLE RUDIMENT

GILL SLITS

MOUTH

MYOTOMES

NERVE CORD

ALIMENTARY CANAL

LARVAL CAUDAL FIN

ANUS

15. Diagram of 3 gill-slit stage based on Specimen 15.

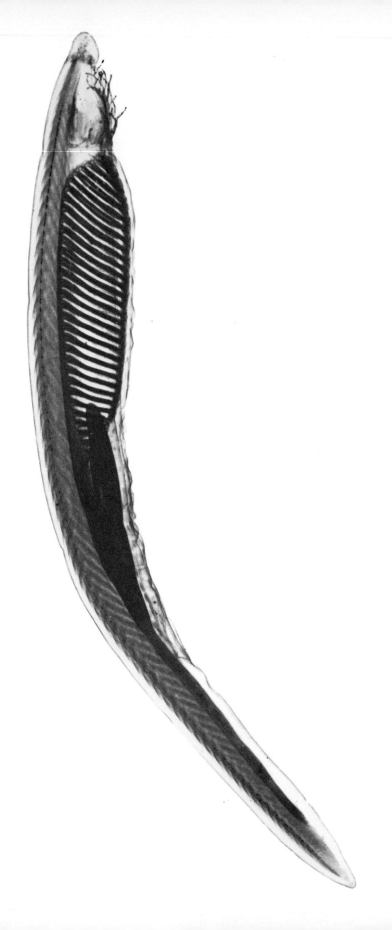

16. **Amphioxus**: immature, E. *mag. 36* ×

11

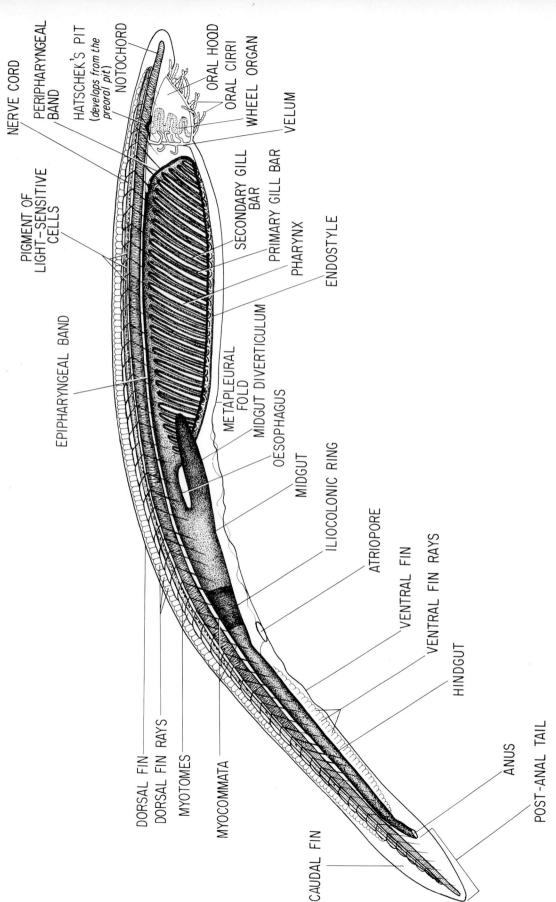

NERVE CORD

PERIPHARYNGEAL BAND

HATSCHEK'S PIT (develops from the preoral pit)

NOTOCHORD

ORAL HOOD

ORAL CIRRI

WHEEL ORGAN

VELUM

PIGMENT OF LIGHT-SENSITIVE CELLS

SECONDARY GILL BAR

PRIMARY GILL BAR

PHARYNX

ENDOSTYLE

EPIPHARYNGEAL BAND

METAPLEURAL FOLD

MIDGUT DIVERTICULUM

OESOPHAGUS

MIDGUT

ILIOCOLONIC RING

ATRIOPORE

VENTRAL FIN

VENTRAL FIN RAYS

HINDGUT

DORSAL FIN

DORSAL FIN RAYS

MYOTOMES

MYOCOMMATA

ANUS

POST-ANAL TAIL

CAUDAL FIN

16. Diagram of Specimen 16.

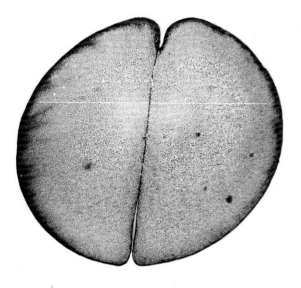

17. Frog: cleavage, 2-cell stage, V.S. *mag. 50×*

18. Frog: cleavage furrows, V.S. *mag. 50×*

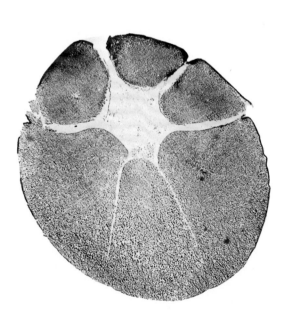

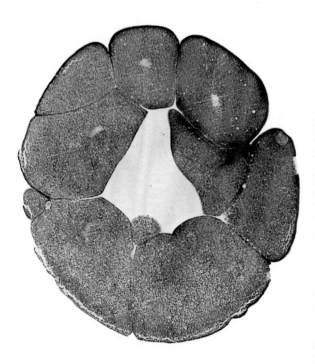

19. Frog: cleavage, 16-cell stage, V.S. *mag. 50×*

20. Frog: cleavage, 24-cell stage, V.S. *mag. 50×*

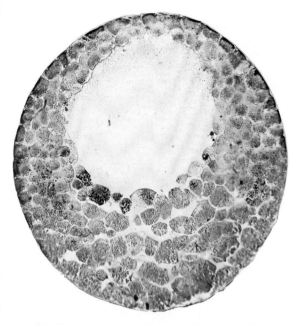

21. **Frog:** cleavage, blastula, V.S. *mag. 45×*

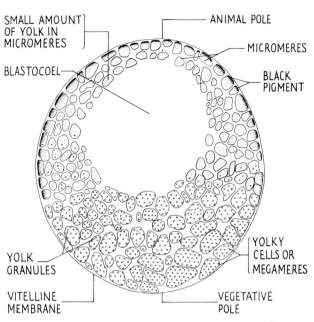

SMALL AMOUNT
OF YOLK IN
MICROMERES

BLASTOCOEL

YOLK
GRANULES

VITELLINE
MEMBRANE

ANIMAL POLE

MICROMERES

BLACK
PIGMENT

YOLKY
CELLS OR
MEGAMERES

VEGETATIVE
POLE

Drawing of Specimen 21

22. **Frog:** early gastrula (dorsal lip), V.S. *mag. 40×*

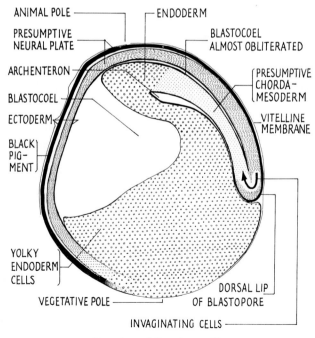

ANIMAL POLE

PRESUMPTIVE
NEURAL PLATE

ARCHENTERON

BLASTOCOEL

ECTODERM

BLACK
PIG-
MENT

YOLKY
ENDODERM
CELLS

VEGETATIVE POLE

ENDODERM

BLASTOCOEL
ALMOST OBLITERATED

PRESUMPTIVE
CHORDA-
MESODERM

VITELLINE
MEMBRANE

DORSAL LIP
OF BLASTOPORE

INVAGINATING CELLS

Drawing of Specimen 22

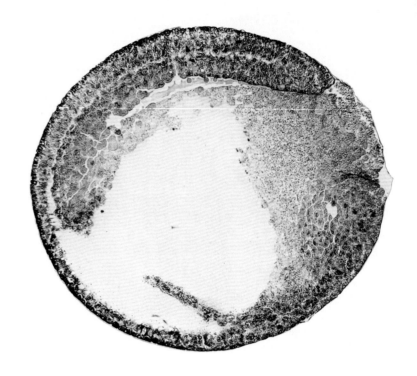

23. **Frog:** later gastrula (yolk plug), V.S. *mag. 60×*

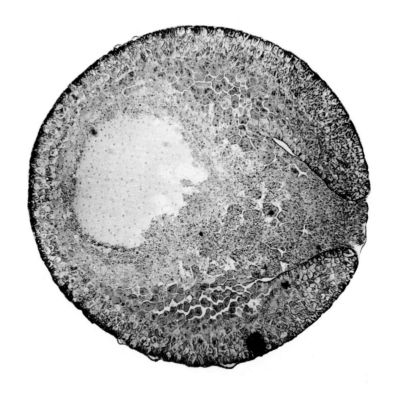

24. **Frog:** later gastrula (yolk plug), H.S. *mag. 60×*

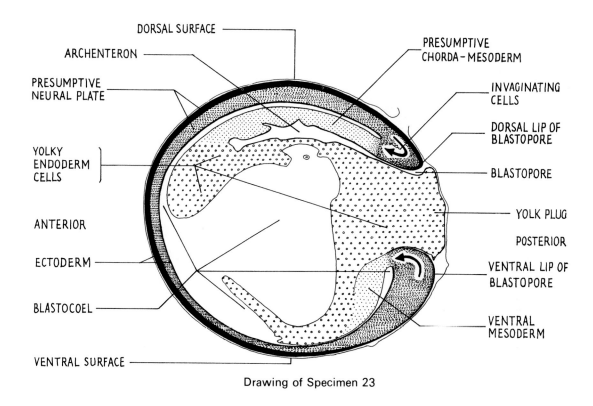

DORSAL SURFACE

ARCHENTERON

PRESUMPTIVE
NEURAL PLATE

PRESUMPTIVE
CHORDA-MESODERM

INVAGINATING
CELLS

DORSAL LIP OF
BLASTOPORE

BLASTOPORE

YOLKY
ENDODERM
CELLS

YOLK PLUG

POSTERIOR

ANTERIOR

ECTODERM

VENTRAL LIP OF
BLASTOPORE

BLASTOCOEL

VENTRAL
MESODERM

VENTRAL SURFACE

Drawing of Specimen 23

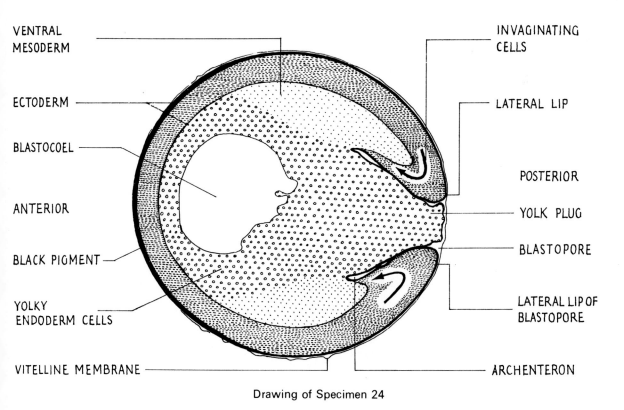

VENTRAL
MESODERM

INVAGINATING
CELLS

ECTODERM

LATERAL LIP

BLASTOCOEL

ANTERIOR

POSTERIOR

YOLK PLUG

BLACK PIGMENT

BLASTOPORE

YOLKY
ENDODERM CELLS

LATERAL LIP OF
BLASTOPORE

VITELLINE MEMBRANE

ARCHENTERON

Drawing of Specimen 24

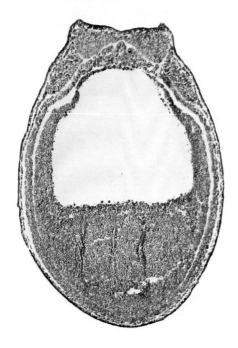

25. **Frog:** neural plate stage, T.S. *mag. 35 ×*

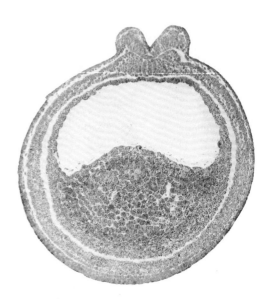

26. **Frog:** neural fold stage, T.S. *mag. 35 ×*

27. **Frog:**
neural tube stage, T.S.
mag. 42 ×

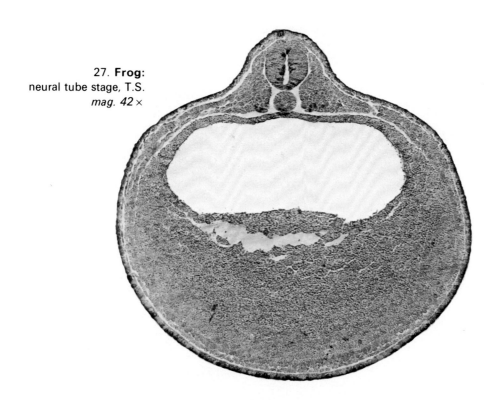

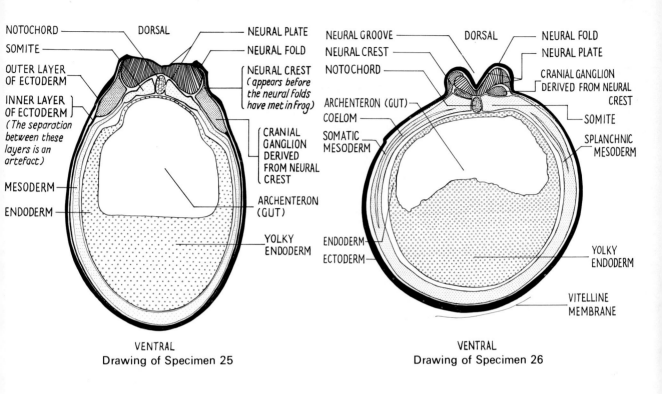

NOTOCHORD — DORSAL — NEURAL PLATE
SOMITE — NEURAL FOLD
OUTER LAYER OF ECTODERM — NEURAL CREST (*appears before the neural folds have met in frog*)
INNER LAYER OF ECTODERM (*The separation between these layers is an artefact*)
CRANIAL GANGLION DERIVED FROM NEURAL CREST
MESODERM — ARCHENTERON (GUT)
ENDODERM — YOLKY ENDODERM

VENTRAL
Drawing of Specimen 25

NEURAL GROOVE — DORSAL — NEURAL FOLD
NEURAL CREST — NEURAL PLATE
NOTOCHORD — CRANIAL GANGLION DERIVED FROM NEURAL CREST
ARCHENTERON (GUT) — SOMITE
COELOM
SOMATIC MESODERM — SPLANCHNIC MESODERM
ENDODERM — YOLKY ENDODERM
ECTODERM
VITELLINE MEMBRANE

VENTRAL
Drawing of Specimen 26

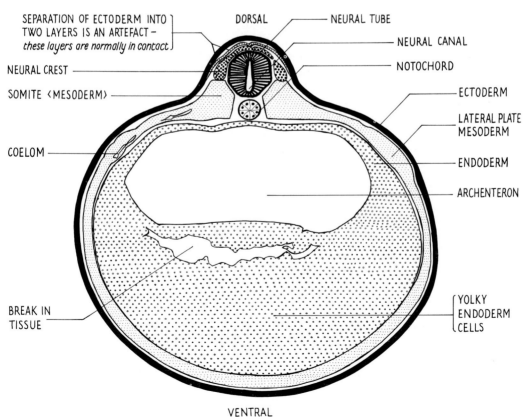

SEPARATION OF ECTODERM INTO TWO LAYERS IS AN ARTEFACT – *these layers are normally in contact*
DORSAL — NEURAL TUBE
NEURAL CANAL
NEURAL CREST — NOTOCHORD
SOMITE ‹MESODERM› — ECTODERM
LATERAL PLATE MESODERM
COELOM — ENDODERM
ARCHENTERON
BREAK IN TISSUE — YOLKY ENDODERM CELLS

VENTRAL
Drawing of Specimen 27

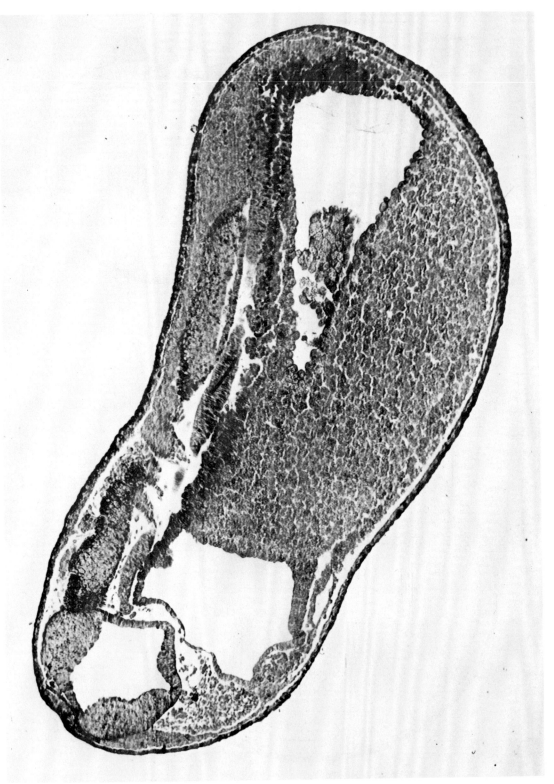

28. **Frog:** *neurula,* V.L.S. *mag. 60* ×

19

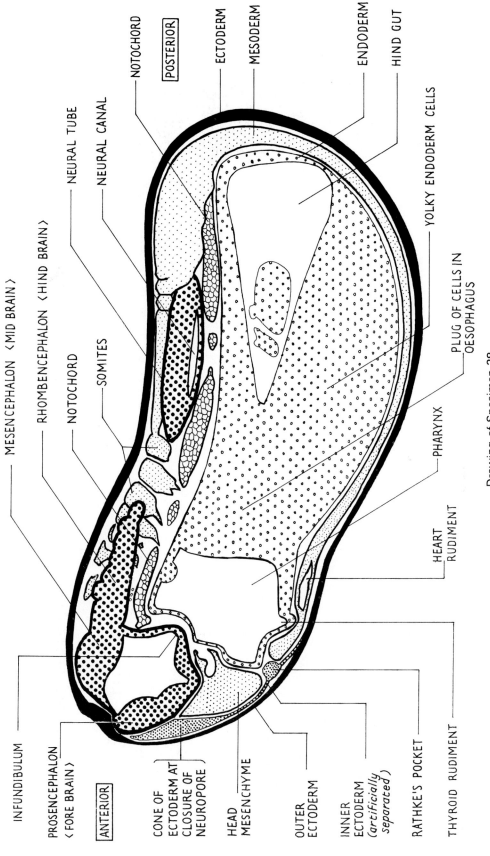

MESENCEPHALON 〈MID BRAIN〉

RHOMBENCEPHALON 〈HIND BRAIN〉

NOTOCHORD

SOMITES

NOTOCHORD

NEURAL TUBE

NEURAL CANAL

POSTERIOR

ECTODERM

MESODERM

ENDODERM

HIND GUT

YOLKY ENDODERM CELLS

PLUG OF CELLS IN OESOPHAGUS

PHARYNX

HEART RUDIMENT

Drawing of Specimen 28

INFUNDIBULUM

PROSENCEPHALON 〈FORE BRAIN〉

ANTERIOR

CONE OF ECTODERM AT CLOSURE OF NEUROPORE

HEAD MESENCHYME

OUTER ECTODERM

INNER ECTODERM (artificially separated)

RATHKE'S POCKET

THYROID RUDIMENT

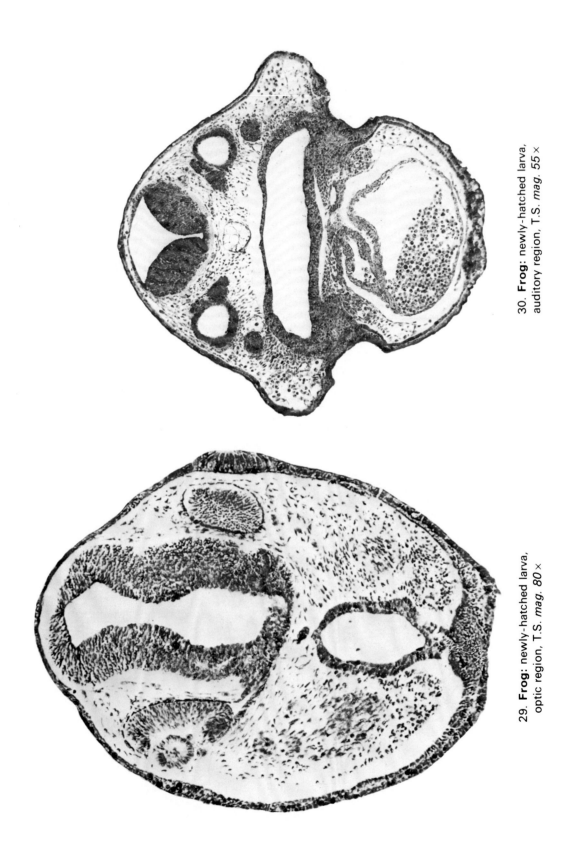

30. **Frog:** newly-hatched larva, auditory region, T.S. *mag. 55 ×*

29. **Frog:** newly-hatched larva, optic region, T.S. *mag. 80 ×*

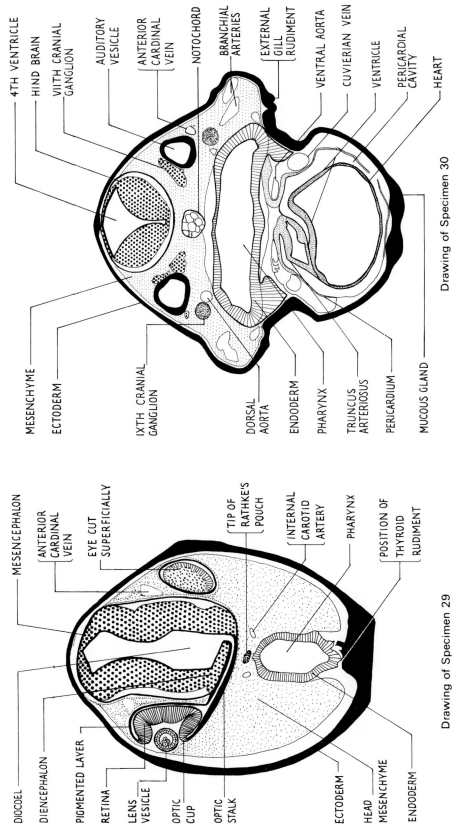

4TH VENTRICLE
HIND BRAIN
VIIth CRANIAL GANGLION
AUDITORY VESICLE
ANTERIOR CARDINAL VEIN
NOTOCHORD
BRANCHIAL ARTERIES
EXTERNAL GILL RUDIMENT
VENTRAL AORTA
CUVIERIAN VEIN
VENTRICLE
PERICARDIAL CAVITY
HEART

MESENCHYME
ECTODERM
IXTH CRANIAL GANGLION
DORSAL AORTA
ENDODERM
PHARYNX
TRUNCUS ARTERIOSUS
PERICARDIUM
MUCOUS GLAND

Drawing of Specimen 30

MESENCEPHALON
ANTERIOR CARDINAL VEIN
EYE CUT SUPERFICIALLY
TIP OF RATHKE'S POUCH
INTERNAL CAROTID ARTERY
PHARYNX
POSITION OF THYROID RUDIMENT

DIOCOEL
DIENCEPHALON
PIGMENTED LAYER
RETINA
LENS VESICLE
OPTIC CUP
OPTIC STALK
ECTODERM
HEAD MESENCHYME
ENDODERM

Drawing of Specimen 29

29 30 31
29 30 31

31. **Frog:** newly-hatched larva,
trunk region, T.S. *mag. 130×*

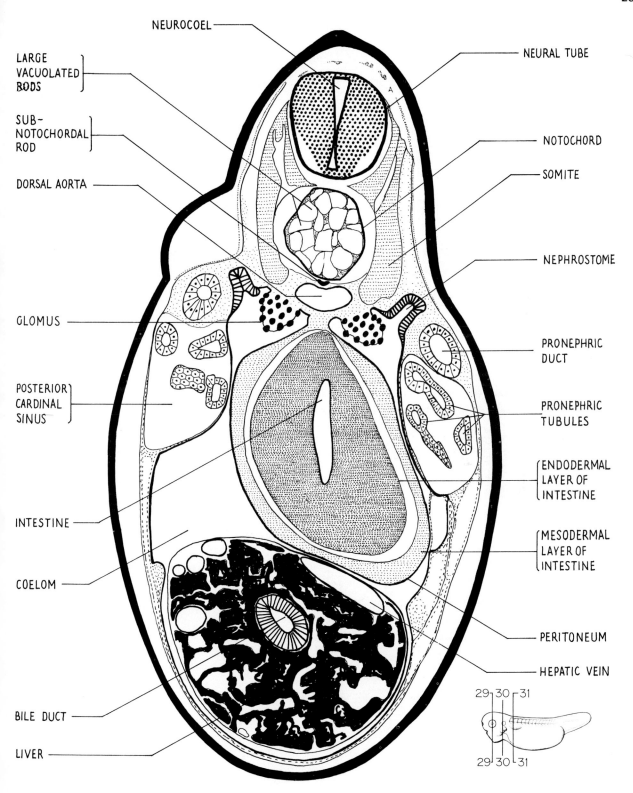

NEUROCOEL

LARGE
VACUOLATED
RODS

SUB-
NOTOCHORDAL
ROD

DORSAL AORTA

GLOMUS

POSTERIOR
CARDINAL
SINUS

INTESTINE

COELOM

BILE DUCT

LIVER

NEURAL TUBE

NOTOCHORD

SOMITE

NEPHROSTOME

PRONEPHRIC
DUCT

PRONEPHRIC
TUBULES

ENDODERMAL
LAYER OF
INTESTINE

MESODERMAL
LAYER OF
INTESTINE

PERITONEUM

HEPATIC VEIN

29 30 31

29 30 31

Drawing of Specimen 31

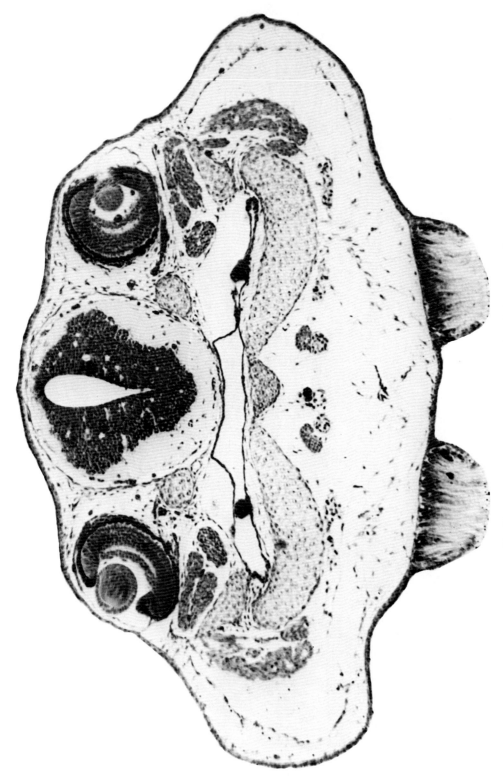

32. **Frog**: external gill larva, optic region, T.S. *mag. 100 ×*

25

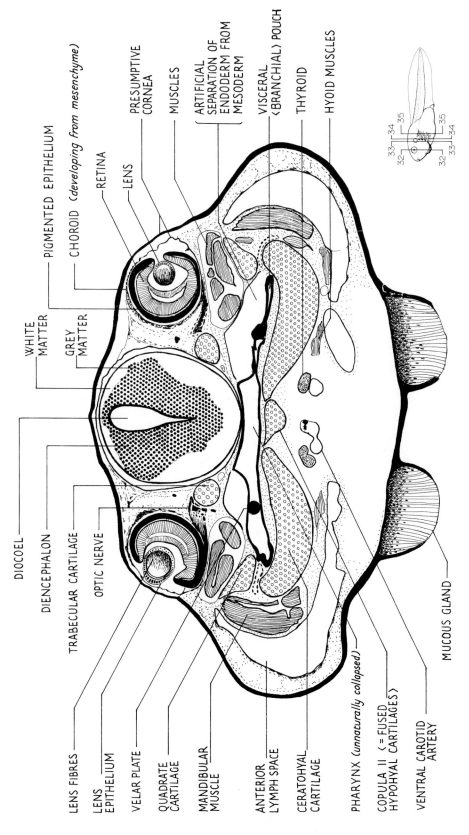

Drawing of Specimen 32

PIGMENTED EPITHELIUM

CHOROID (developing from mesenchyme)

RETINA

LENS

PRESUMPTIVE CORNEA

MUSCLES

ARTIFICIAL SEPARATION OF ENDODERM FROM MESODERM

VISCERAL (BRANCHIAL) POUCH

THYROID

HYOID MUSCLES

WHITE MATTER

GREY MATTER

DIOCOEL

DIENCEPHALON

TRABECULAR CARTILAGE

OPTIC NERVE

LENS FIBRES

LENS EPITHELIUM

VELAR PLATE

QUADRATE CARTILAGE

MANDIBULAR MUSCLE

ANTERIOR LYMPH SPACE

CERATOHYAL CARTILAGE

PHARYNX (unnaturally collapsed)

COPULA II (= FUSED HYPOHYAL CARTILAGES)

VENTRAL CAROTID ARTERY

MUCOUS GLAND

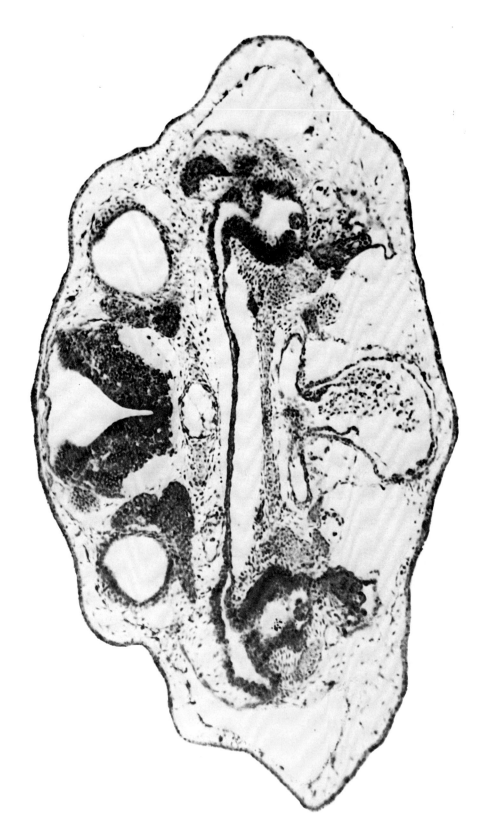

33. **Frog**: external gill larva, auditory region, T.S. *mag. 100 ×*

27

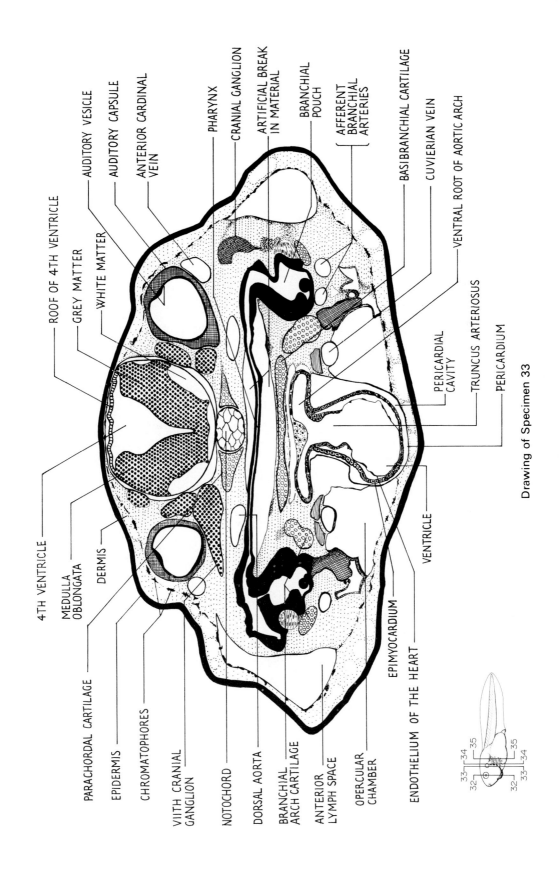

Drawing of Specimen 33

PARACHORDAL CARTILAGE

EPIDERMIS

CHROMATOPHORES

VIITH CRANIAL GANGLION

NOTOCHORD

DORSAL AORTA

BRANCHIAL ARCH CARTILAGE

ANTERIOR LYMPH SPACE

OPERCULAR CHAMBER

ENDOTHELIUM OF THE HEART

EPIMYOCARDIUM

VENTRICLE

4TH VENTRICLE

MEDULLA OBLONGATA

DERMIS

ROOF OF 4TH VENTRICLE

GREY MATTER

WHITE MATTER

AUDITORY VESICLE

AUDITORY CAPSULE

ANTERIOR CARDINAL VEIN

PHARYNX

CRANIAL GANGLION

ARTIFICIAL BREAK IN MATERIAL

BRANCHIAL POUCH

AFFERENT BRANCHIAL ARTERIES

BASIBRANCHIAL CARTILAGE

CUVIERIAN VEIN

VENTRAL ROOT OF AORTIC ARCH

PERICARDIAL CAVITY

TRUNCUS ARTERIOSUS

PERICARDIUM

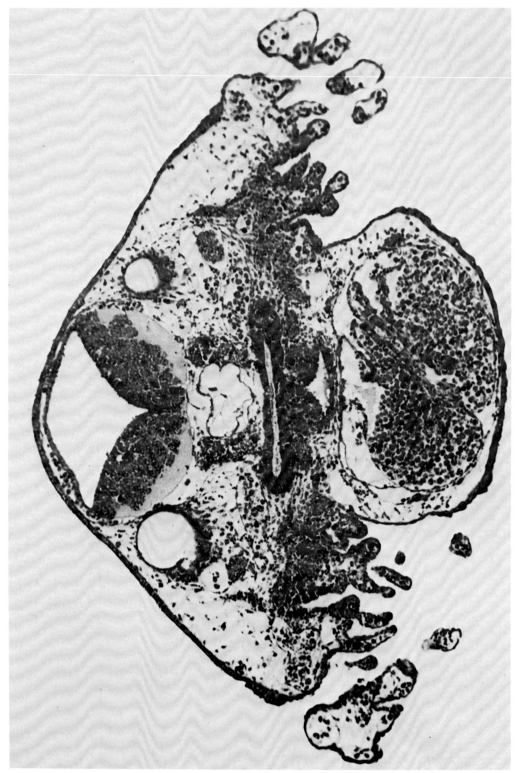

34. **Frog**: external gill larva, heart and gill region, T.S. *mag. 120* ×

29

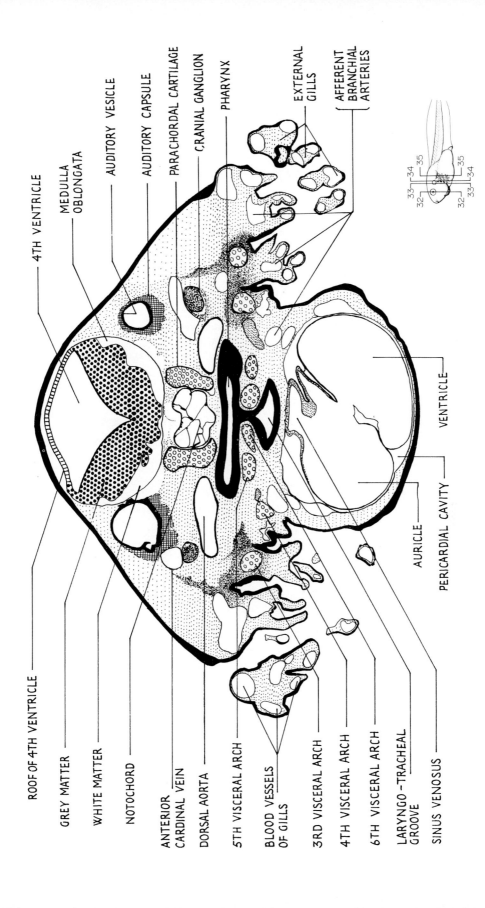

Drawing of Specimen 34

ROOF OF 4TH VENTRICLE

GREY MATTER

WHITE MATTER

NOTOCHORD

ANTERIOR CARDINAL VEIN

DORSAL AORTA

5TH VISCERAL ARCH

BLOOD VESSELS OF GILLS

3RD VISCERAL ARCH

4TH VISCERAL ARCH

6TH VISCERAL ARCH

LARYNGO-TRACHEAL GROOVE

SINUS VENOSUS

4TH VENTRICLE

MEDULLA OBLONGATA

AUDITORY VESICLE

AUDITORY CAPSULE

PARACHORDAL CARTILAGE

CRANIAL GANGLION

PHARYNX

EXTERNAL GILLS

AFFERENT BRANCHIAL ARTERIES

VENTRICLE

AURICLE

PERICARDIAL CAVITY

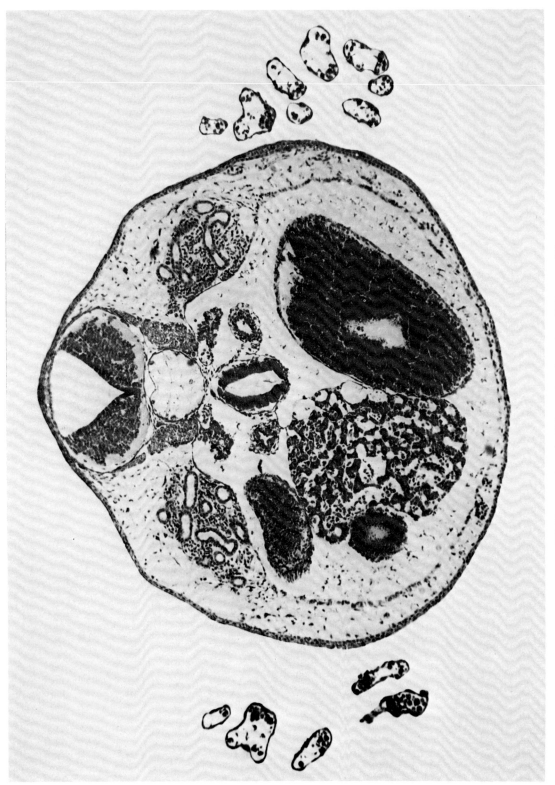

35. **Frog**: external gill larva, trunk region, T.S. *mag. 80* ×

31

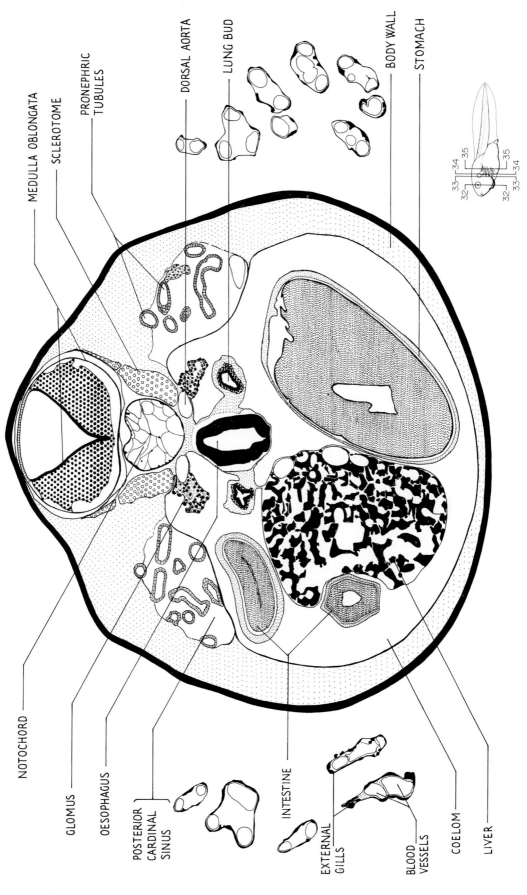

MEDULLA OBLONGATA

SCLEROTOME

PRONEPHRIC TUBULES

DORSAL AORTA

LUNG BUD

BODY WALL

STOMACH

NOTOCHORD

GLOMUS

OESOPHAGUS

POSTERIOR CARDINAL SINUS

INTESTINE

EXTERNAL GILLS

BLOOD VESSELS

COELOM

LIVER

Drawing of Specimen 35

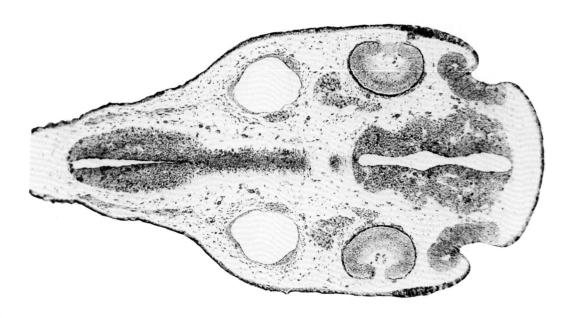

36. **Frog:** external gill larva, head region, H.L.S. *mag. 85 ×*

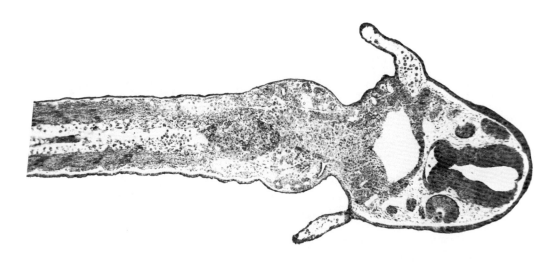

37. **Frog:** external gill larva, trunk region, H.L.S. *mag. 50 ×*

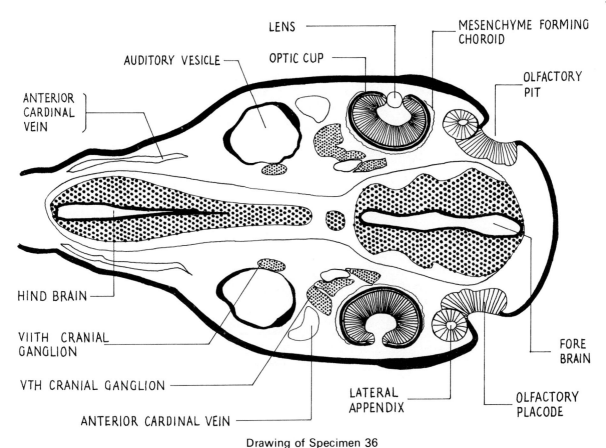

LENS

MESENCHYME FORMING
CHOROID

AUDITORY VESICLE

OPTIC CUP

OLFACTORY
PIT

ANTERIOR
CARDINAL
VEIN

HIND BRAIN

VIITH CRANIAL
GANGLION

VTH CRANIAL GANGLION

ANTERIOR CARDINAL VEIN

LATERAL
APPENDIX

FORE
BRAIN

OLFACTORY
PLACODE

Drawing of Specimen 36

OESOPHAGUS CLOSED
BY PLUG OF CELLS

EXTERNAL
GILL

PHARYNX

DORSAL AORTA

PRONEPHRIC BULGE

DORSAL AORTA

MYOTOMES

STOMACH

INFUNDIBULUM

NOTOCHORD

OLFACTORY
ORGAN

TAIL

COELOM

POSTERIOR CARDINAL SINUS

NEPHROSTOME

PRONEPHRIC TUBULES

AORTIC ARCHES

TELOCOEL

TELENCEPHALON

VISCERAL ARCHES

VTH, VIITH & IXTH
CRANIAL GANGLIA

OPTIC CUP

LENS

Drawing of Specimen 37

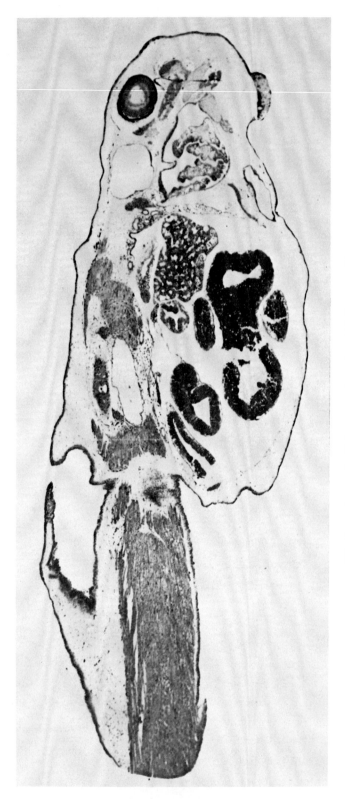

38. **Frog:** internal gill larva, trunk region, V.L.S. *mag. 40* ×

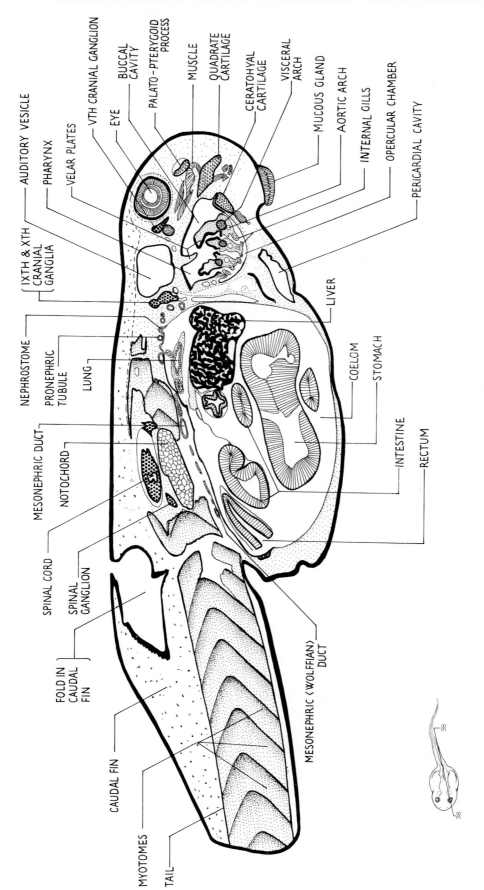

Drawing of Specimen 38

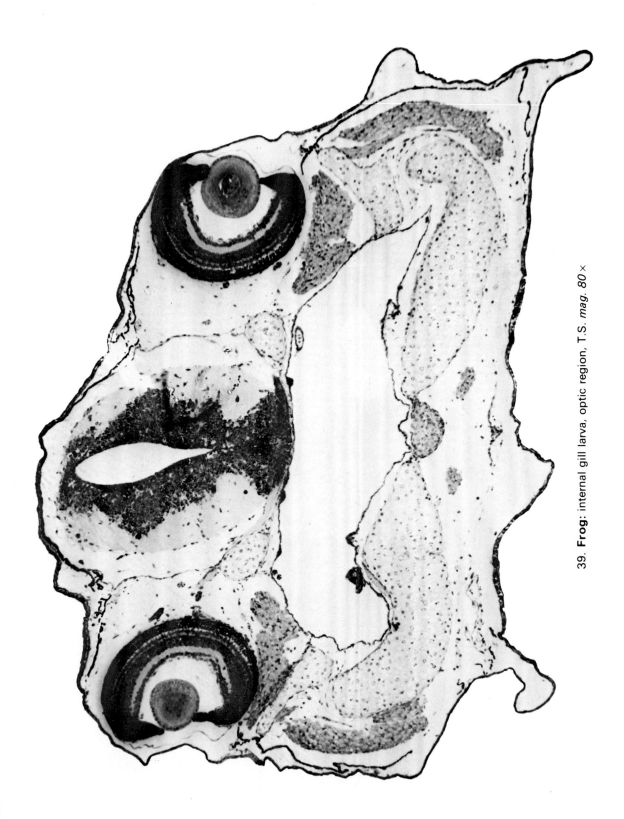

39. **Frog**: internal gill larva, optic region, T.S. *mag. 80×*

37

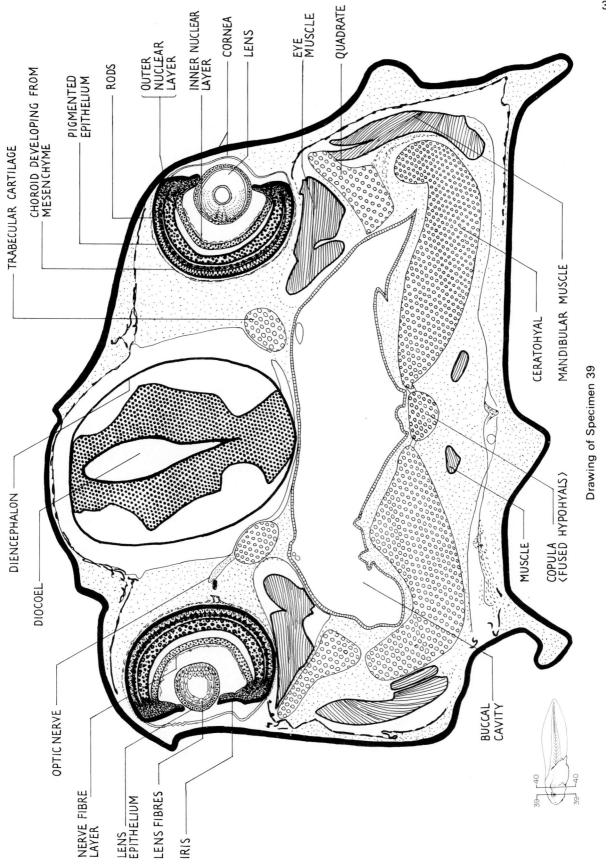

TRABECULAR CARTILAGE

CHOROID DEVELOPING FROM MESENCHYME

PIGMENTED EPITHELIUM

RODS

OUTER NUCLEAR LAYER

INNER NUCLEAR LAYER

CORNEA

LENS

EYE MUSCLE

QUADRATE

DIENCEPHALON

DIOCOEL

OPTIC NERVE

NERVE FIBRE LAYER

LENS EPITHELIUM

LENS FIBRES

IRIS

BUCCAL CAVITY

MUSCLE

COPULA ⟨FUSED HYPOHYALS⟩

CERATOHYAL

MANDIBULAR MUSCLE

Drawing of Specimen 39

39 40

39 40

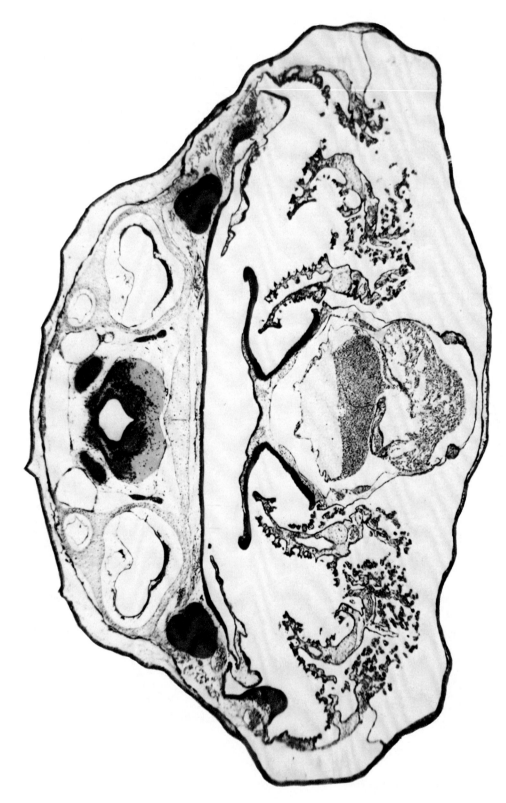

40. **Frog**: internal gill larva, gill region, T.S. *mag. 45* ×

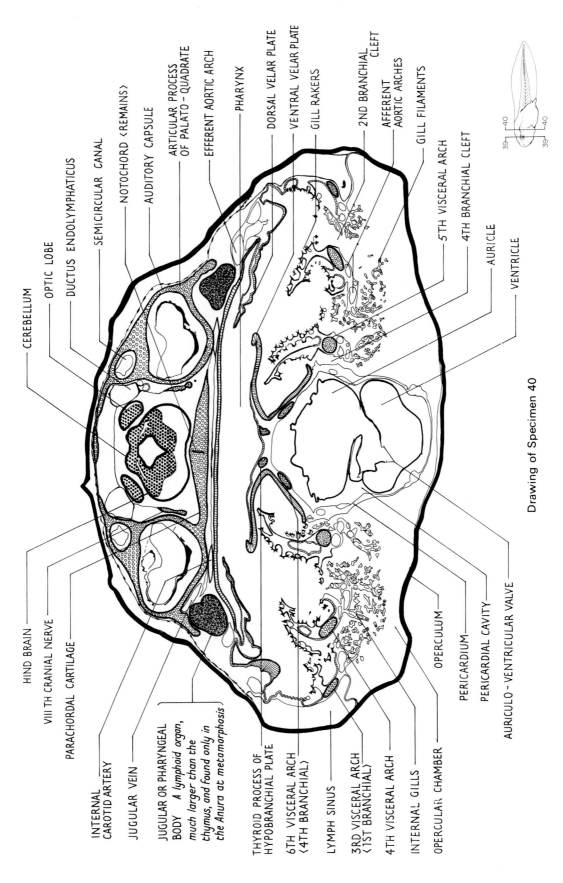

CEREBELLUM

OPTIC LOBE

DUCTUS ENDOLYMPHATICUS

SEMICIRCULAR CANAL

NOTOCHORD ⟨REMAINS⟩

AUDITORY CAPSULE

ARTICULAR PROCESS
OF PALATO - QUADRATE

EFFERENT AORTIC ARCH

PHARYNX

DORSAL VELAR PLATE

VENTRAL VELAR PLATE

GILL RAKERS

2ND BRANCHIAL
CLEFT

AFFERENT
AORTIC ARCHES

GILL FILAMENTS

5TH VISCERAL ARCH

4TH BRANCHIAL CLEFT

AURICLE

VENTRICLE

HIND BRAIN

VIIITH CRANIAL NERVE

PARACHORDAL CARTILAGE

INTERNAL
CAROTID ARTERY

JUGULAR VEIN

JUGULAR OR PHARYNGEAL
BODY *A lymphoid organ,
much larger than the
thymus, and found only in
the Anura at metamorphosis*

THYROID PROCESS OF
HYPOBRANCHIAL PLATE

6TH VISCERAL ARCH
⟨4TH BRANCHIAL⟩

LYMPH SINUS

3RD VISCERAL ARCH
⟨1ST BRANCHIAL⟩

4TH VISCERAL ARCH

INTERNAL GILLS

OPERCULAR CHAMBER

OPERCULUM

PERICARDIUM

PERICARDIAL CAVITY

AURICULO - VENTRICULAR VALVE

Drawing of Specimen 40

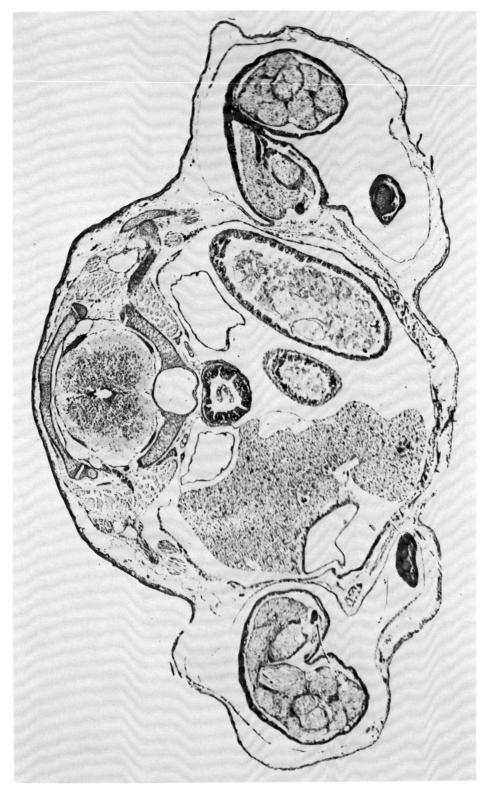

41. **Frog**: 19-mm tadpole, forelimb region, T.S. *mag. 35* ×

41

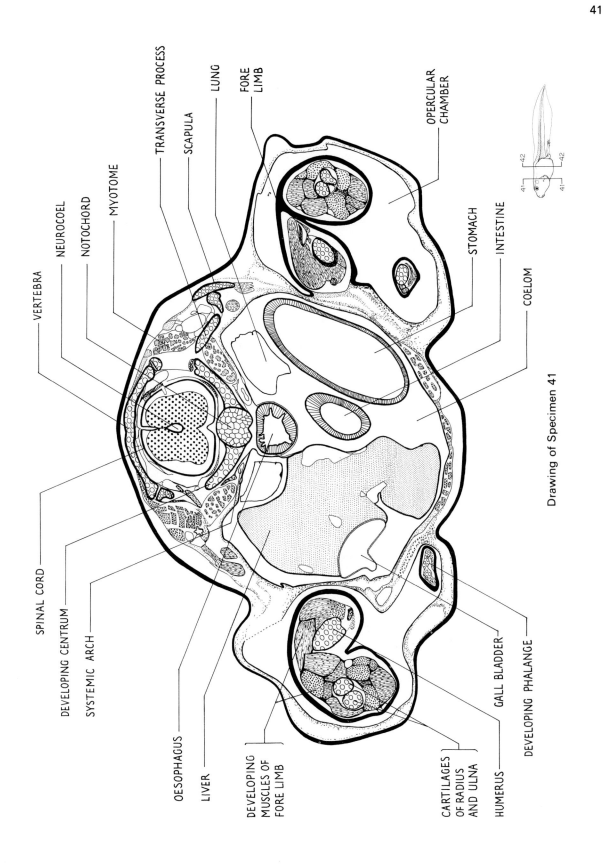

Drawing of Specimen 41

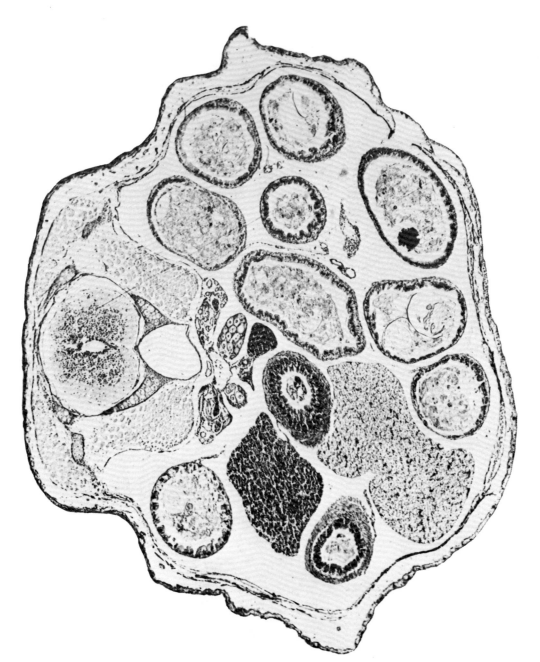

42. **Frog:** 19-mm tadpole, trunk region, T.S. *mag. 40* ×

43

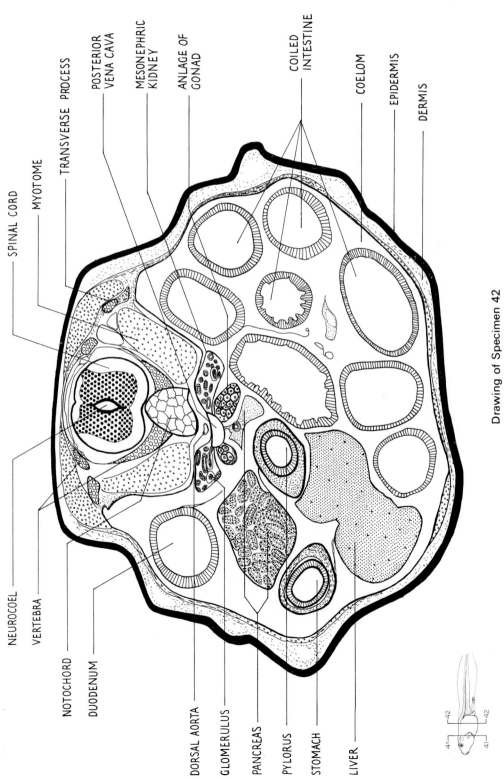

SPINAL CORD

MYOTOME

TRANSVERSE PROCESS

POSTERIOR VENA CAVA

MESONEPHRIC KIDNEY

ANLAGE OF GONAD

COILED INTESTINE

COELOM

EPIDERMIS

DERMIS

NEUROCOEL

VERTEBRA

NOTOCHORD

DUODENUM

DORSAL AORTA

GLOMERULUS

PANCREAS

PYLORUS

STOMACH

LIVER

Drawing of Specimen 42

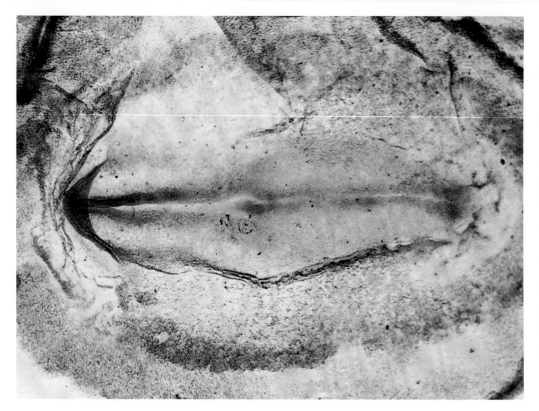

44. **Chick**: blastoderm, head-folded stage, E. *mag.* 25 ×

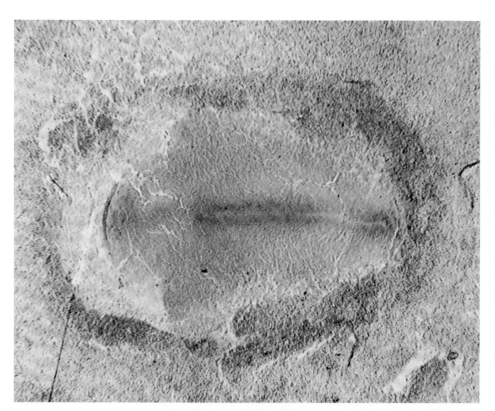

43. **Chick**: blastoderm, head-process stage, E. *mag.* 25 ×

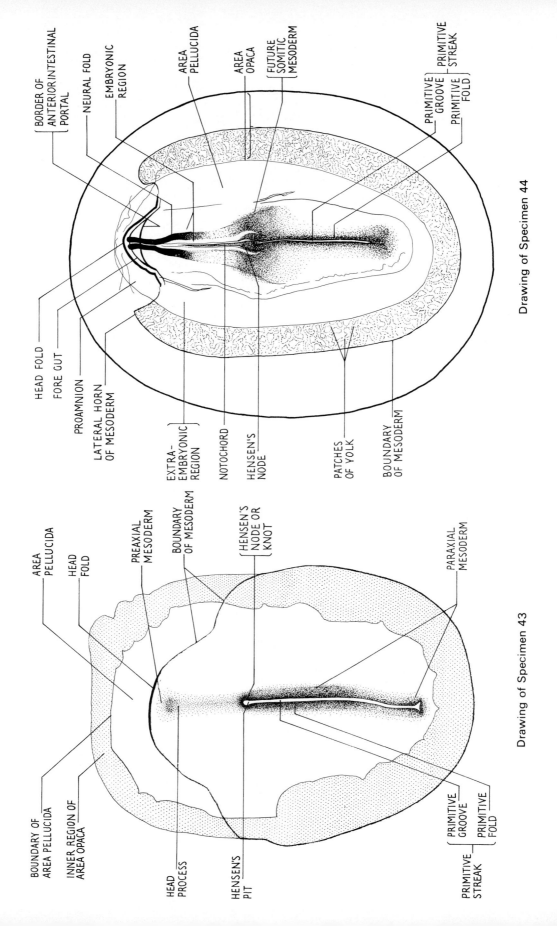

45

BORDER OF
ANTERIOR INTESTINAL
PORTAL

NEURAL FOLD

EMBRYONIC
REGION

AREA
PELLUCIDA

AREA
OPACA

FUTURE
SOMITIC
MESODERM

PRIMITIVE
GROOVE

PRIMITIVE
STREAK

PRIMITIVE
FOLD

HEAD FOLD

FORE GUT

PROAMNION

LATERAL HORN
OF MESODERM

EXTRA-
EMBRYONIC
REGION

NOTOCHORD

HENSEN'S
NODE

PATCHES
OF YOLK

BOUNDARY
OF MESODERM

Drawing of Specimen 44

AREA
PELLUCIDA

HEAD
FOLD

PREAXIAL
MESODERM

BOUNDARY
OF MESODERM

HENSEN'S
NODE OR
KNOT

PARAXIAL
MESODERM

BOUNDARY OF
AREA PELLUCIDA

INNER REGION OF
AREA OPACA

HEAD
PROCESS

HENSEN'S
PIT

PRIMITIVE
GROOVE

PRIMITIVE
FOLD

PRIMITIVE
STREAK

Drawing of Specimen 43

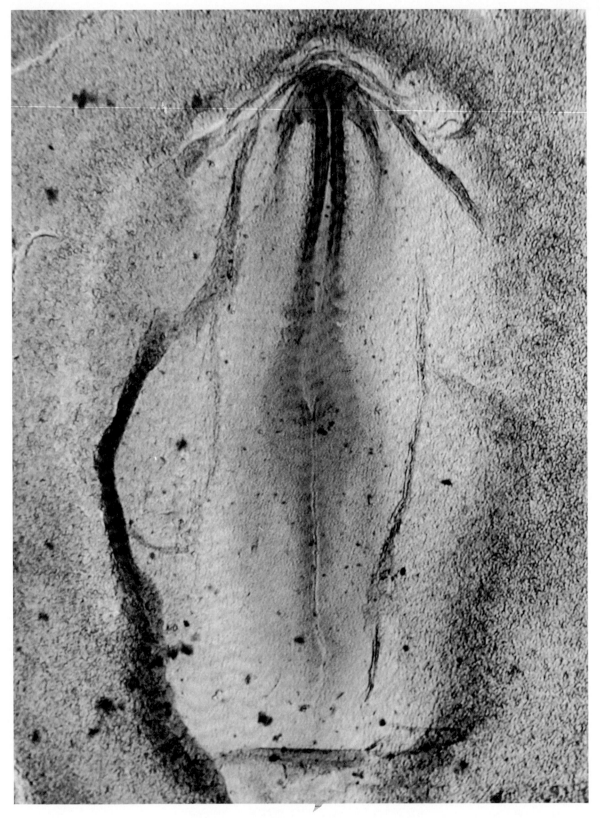

45. **Chick:** blastoderm, 3-somite, E. *mag. 40×*

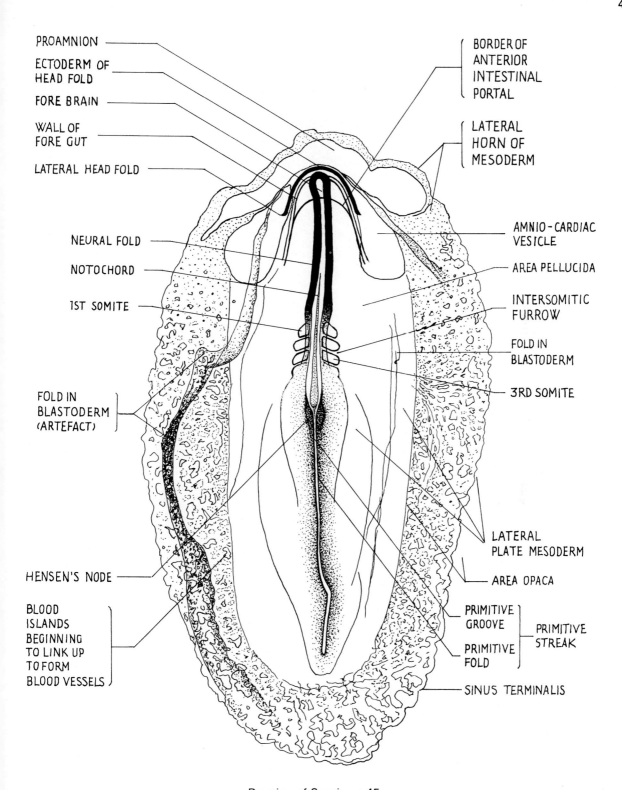

PROAMNION

ECTODERM OF
HEAD FOLD

FORE BRAIN

WALL OF
FORE GUT

LATERAL HEAD FOLD

NEURAL FOLD

NOTOCHORD

1ST SOMITE

FOLD IN
BLASTODERM
(ARTEFACT)

HENSEN'S NODE

BLOOD
ISLANDS
BEGINNING
TO LINK UP
TO FORM
BLOOD VESSELS

BORDER OF
ANTERIOR
INTESTINAL
PORTAL

LATERAL
HORN OF
MESODERM

AMNIO-CARDIAC
VESICLE

AREA PELLUCIDA

INTERSOMITIC
FURROW

FOLD IN
BLASTODERM

3RD SOMITE

LATERAL
PLATE MESODERM

AREA OPACA

PRIMITIVE
GROOVE

PRIMITIVE
STREAK

PRIMITIVE
FOLD

SINUS TERMINALIS

Drawing of Specimen 45

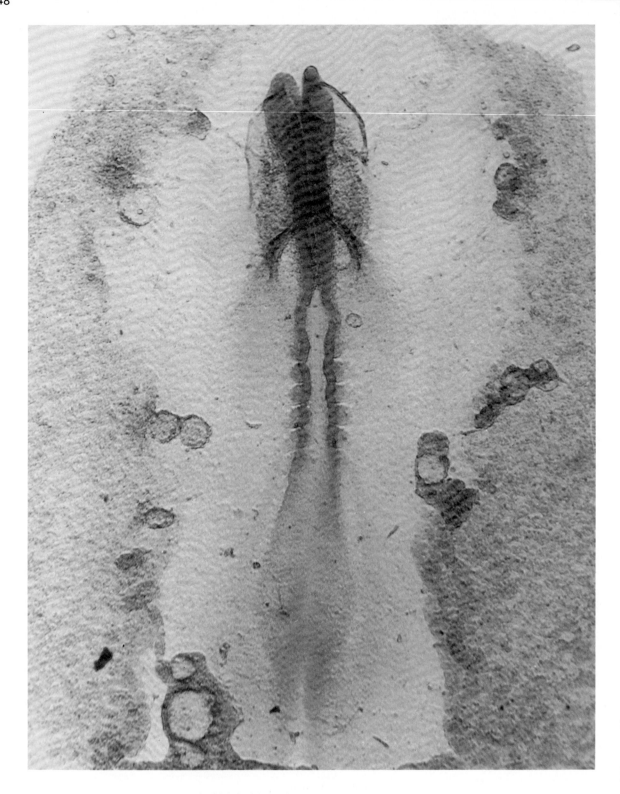

46. **Chick:** blastoderm, 6-somite, E. *mag. 40 ×*

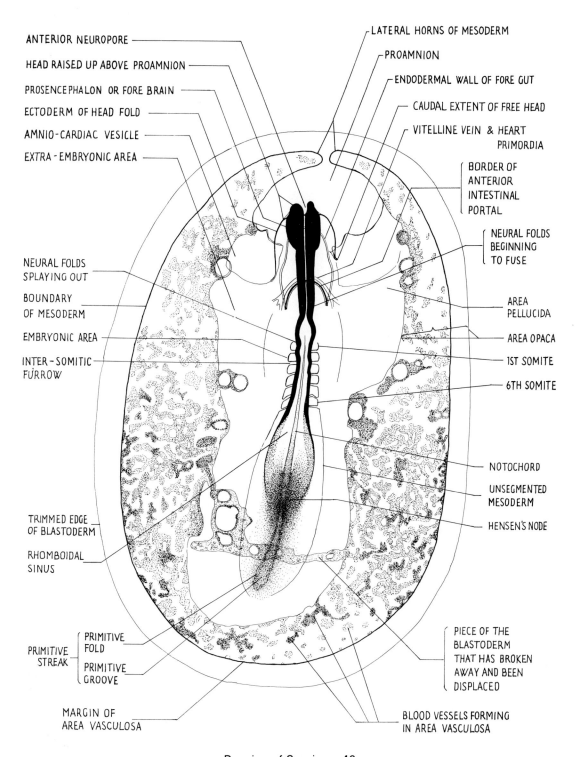

ANTERIOR NEUROPORE

HEAD RAISED UP ABOVE PROAMNION

PROSENCEPHALON OR FORE BRAIN

ECTODERM OF HEAD FOLD

AMNIO-CARDIAC VESICLE

EXTRA-EMBRYONIC AREA

LATERAL HORNS OF MESODERM

PROAMNION

ENDODERMAL WALL OF FORE GUT

CAUDAL EXTENT OF FREE HEAD

VITELLINE VEIN & HEART PRIMORDIA

BORDER OF ANTERIOR INTESTINAL PORTAL

NEURAL FOLDS BEGINNING TO FUSE

NEURAL FOLDS SPLAYING OUT

BOUNDARY OF MESODERM

EMBRYONIC AREA

INTER-SOMITIC FURROW

AREA PELLUCIDA

AREA OPACA

1ST SOMITE

6TH SOMITE

NOTOCHORD

UNSEGMENTED MESODERM

HENSEN'S NODE

TRIMMED EDGE OF BLASTODERM

RHOMBOIDAL SINUS

PRIMITIVE STREAK

PRIMITIVE FOLD

PRIMITIVE GROOVE

PIECE OF THE BLASTODERM THAT HAS BROKEN AWAY AND BEEN DISPLACED

MARGIN OF AREA VASCULOSA

BLOOD VESSELS FORMING IN AREA VASCULOSA

Drawing of Specimen 46

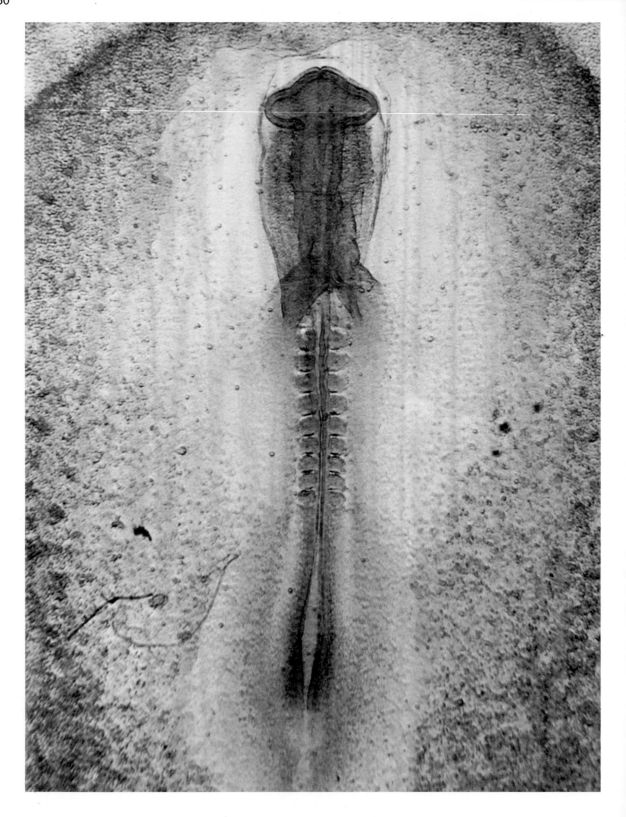

47. **Chick:** blastoderm, 10-somite, E. *mag. 45 ×*

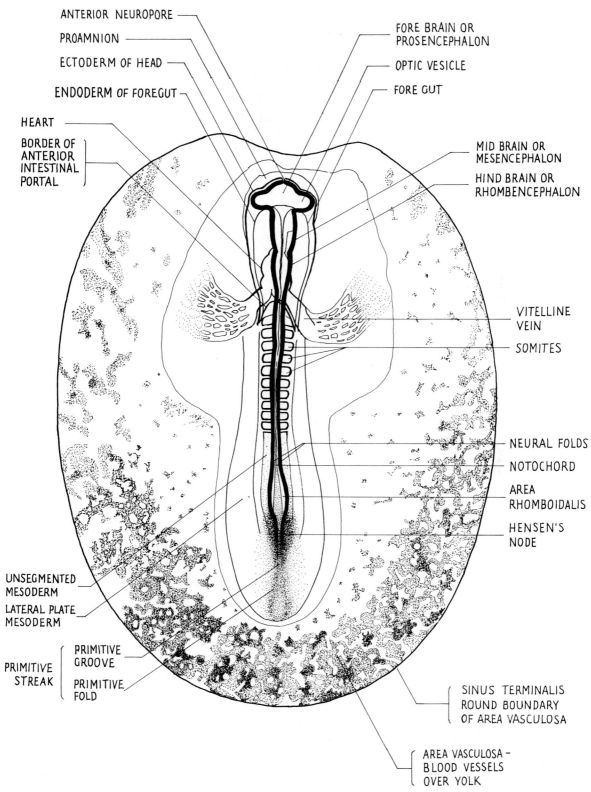

ANTERIOR NEUROPORE

PROAMNION

ECTODERM OF HEAD

ENDODERM OF FOREGUT

HEART

BORDER OF
ANTERIOR
INTESTINAL
PORTAL

FORE BRAIN OR
PROSENCEPHALON

OPTIC VESICLE

FORE GUT

MID BRAIN OR
MESENCEPHALON

HIND BRAIN OR
RHOMBENCEPHALON

VITELLINE
VEIN

SOMITES

NEURAL FOLDS

NOTOCHORD

AREA
RHOMBOIDALIS

HENSEN'S
NODE

UNSEGMENTED
MESODERM

LATERAL PLATE
MESODERM

PRIMITIVE
STREAK

PRIMITIVE
GROOVE

PRIMITIVE
FOLD

SINUS TERMINALIS
ROUND BOUNDARY
OF AREA VASCULOSA

AREA VASCULOSA –
BLOOD VESSELS
OVER YOLK

(Drawn from ventral aspect; photograph is of dorsal aspect)

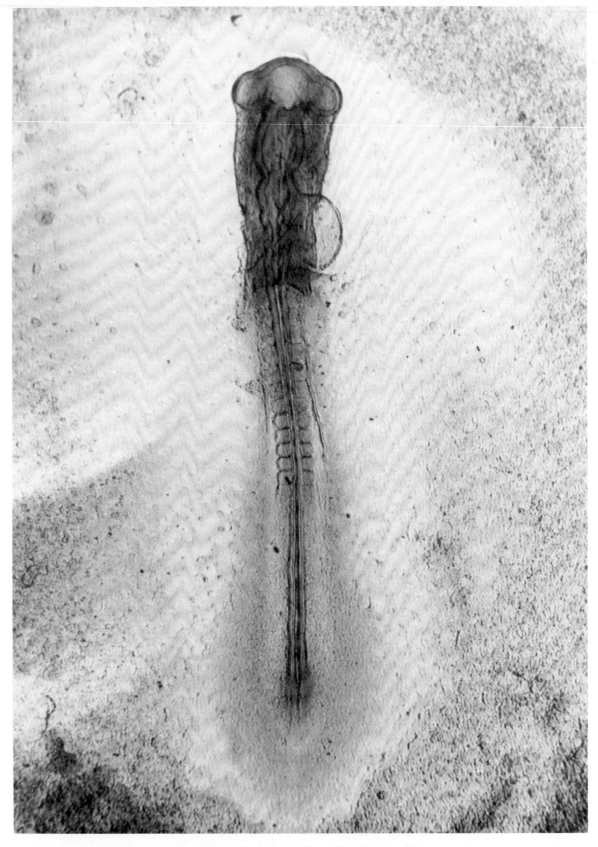

48. **Chick:** blastoderm, 13-somite, E. *mag. 35 ×*

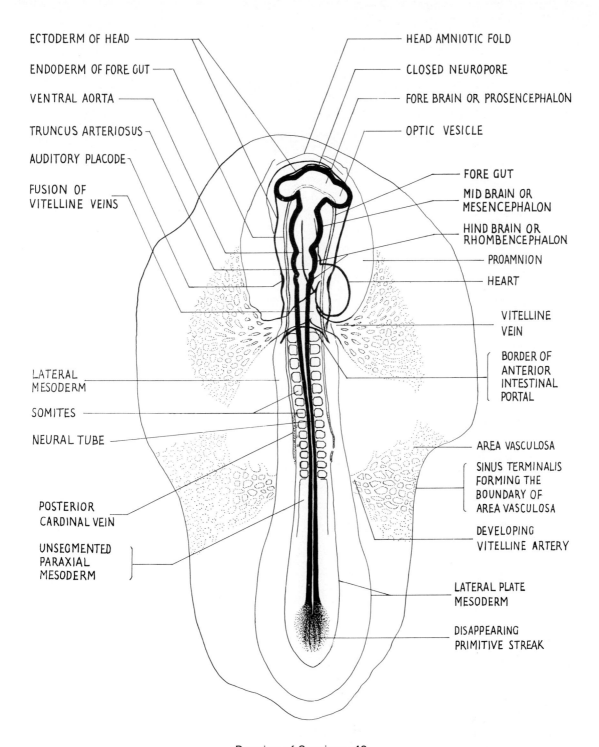

ECTODERM OF HEAD

ENDODERM OF FORE GUT

VENTRAL AORTA

TRUNCUS ARTERIOSUS

AUDITORY PLACODE

FUSION OF
VITELLINE VEINS

LATERAL
MESODERM

SOMITES

NEURAL TUBE

POSTERIOR
CARDINAL VEIN

UNSEGMENTED
PARAXIAL
MESODERM

HEAD AMNIOTIC FOLD

CLOSED NEUROPORE

FORE BRAIN OR PROSENCEPHALON

OPTIC VESICLE

FORE GUT

MID BRAIN OR
MESENCEPHALON

HIND BRAIN OR
RHOMBENCEPHALON

PROAMNION

HEART

VITELLINE
VEIN

BORDER OF
ANTERIOR
INTESTINAL
PORTAL

AREA VASCULOSA

SINUS TERMINALIS
FORMING THE
BOUNDARY OF
AREA VASCULOSA

DEVELOPING
VITELLINE ARTERY

LATERAL PLATE
MESODERM

DISAPPEARING
PRIMITIVE STREAK

Drawing of Specimen 48

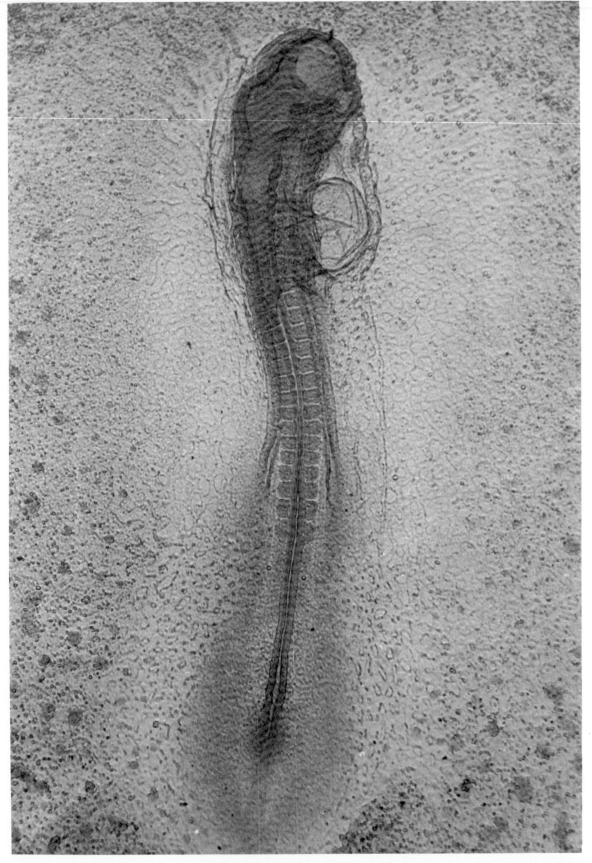

49. **Chick:** blastoderm, 17-somite, E. *mag. 30×*

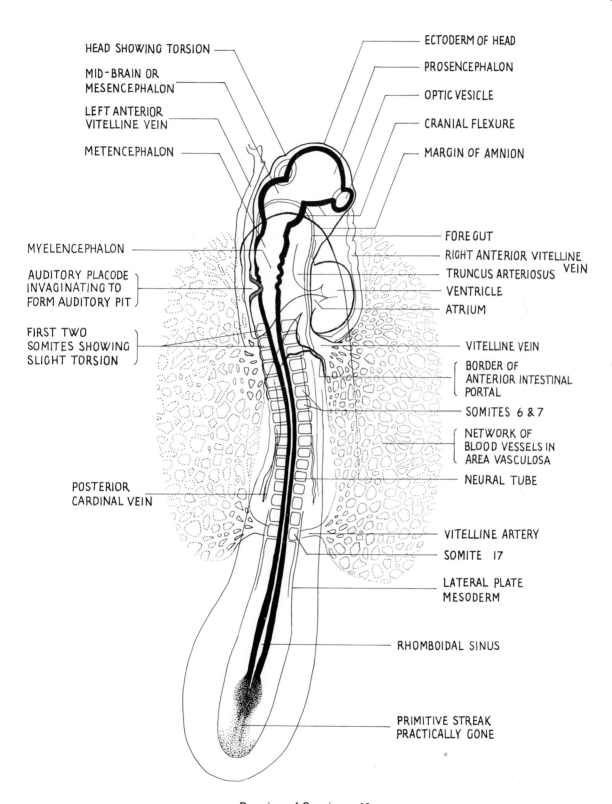

HEAD SHOWING TORSION

MID-BRAIN OR MESENCEPHALON

LEFT ANTERIOR VITELLINE VEIN

METENCEPHALON

ECTODERM OF HEAD

PROSENCEPHALON

OPTIC VESICLE

CRANIAL FLEXURE

MARGIN OF AMNION

MYELENCEPHALON

AUDITORY PLACODE INVAGINATING TO FORM AUDITORY PIT

FIRST TWO SOMITES SHOWING SLIGHT TORSION

FORE GUT

RIGHT ANTERIOR VITELLINE VEIN

TRUNCUS ARTERIOSUS

VENTRICLE

ATRIUM

VITELLINE VEIN

BORDER OF ANTERIOR INTESTINAL PORTAL

SOMITES 6 & 7

NETWORK OF BLOOD VESSELS IN AREA VASCULOSA

NEURAL TUBE

POSTERIOR CARDINAL VEIN

VITELLINE ARTERY

SOMITE 17

LATERAL PLATE MESODERM

RHOMBOIDAL SINUS

PRIMITIVE STREAK PRACTICALLY GONE

Drawing of Specimen 49

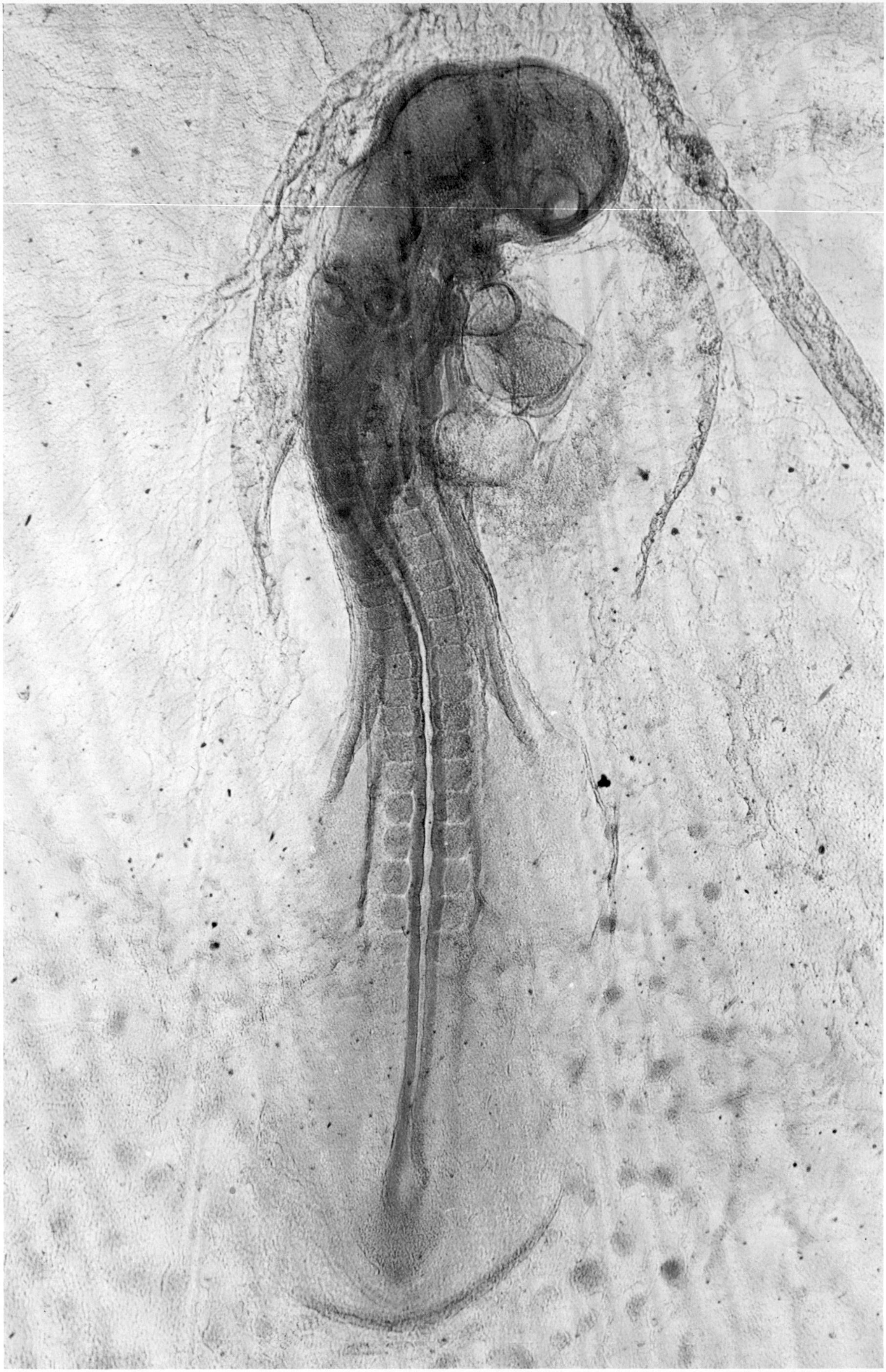

57

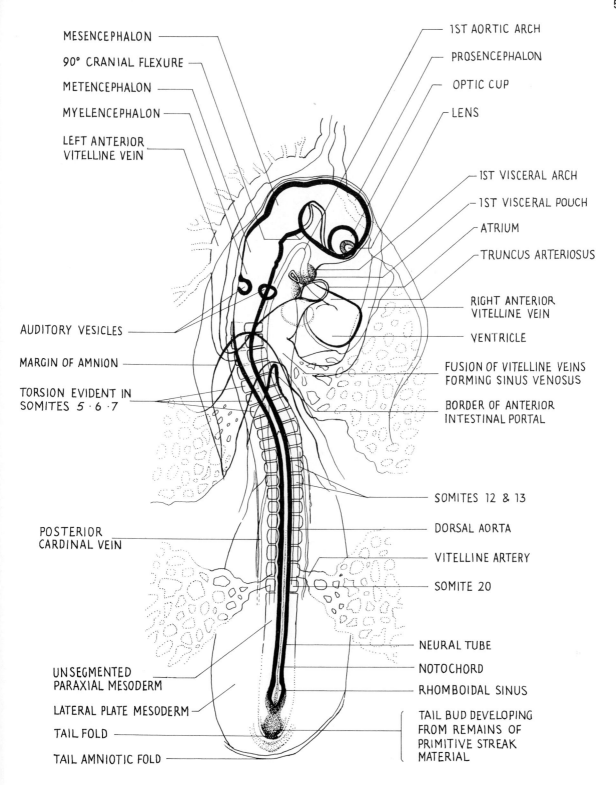

MESENCEPHALON

90° CRANIAL FLEXURE

METENCEPHALON

MYELENCEPHALON

LEFT ANTERIOR
VITELLINE VEIN

1ST AORTIC ARCH

PROSENCEPHALON

OPTIC CUP

LENS

1ST VISCERAL ARCH

1ST VISCERAL POUCH

ATRIUM

TRUNCUS ARTERIOSUS

RIGHT ANTERIOR
VITELLINE VEIN

VENTRICLE

AUDITORY VESICLES

MARGIN OF AMNION

TORSION EVIDENT IN
SOMITES 5 · 6 · 7

FUSION OF VITELLINE VEINS
FORMING SINUS VENOSUS

BORDER OF ANTERIOR
INTESTINAL PORTAL

SOMITES 12 & 13

DORSAL AORTA

VITELLINE ARTERY

SOMITE 20

POSTERIOR
CARDINAL VEIN

NEURAL TUBE

NOTOCHORD

RHOMBOIDAL SINUS

UNSEGMENTED
PARAXIAL MESODERM

LATERAL PLATE MESODERM

TAIL FOLD

TAIL AMNIOTIC FOLD

TAIL BUD DEVELOPING
FROM REMAINS OF
PRIMITIVE STREAK
MATERIAL

Drawing of Specimen 50

(*Left*) 50. **Chick:** blastoderm, 20-somite, E. *mag. 40×*

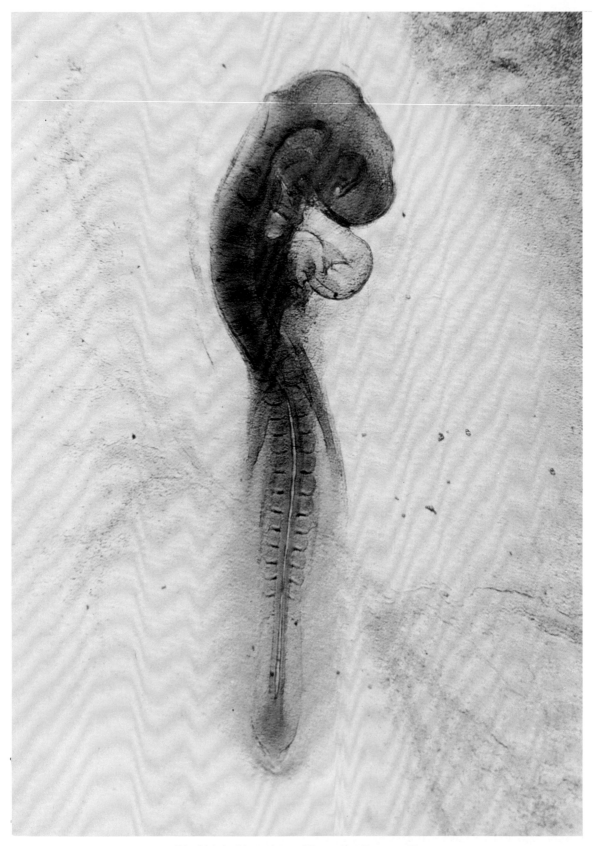

51. **Chick:** blastoderm, 25-somite, E. *mag. 45 ×*

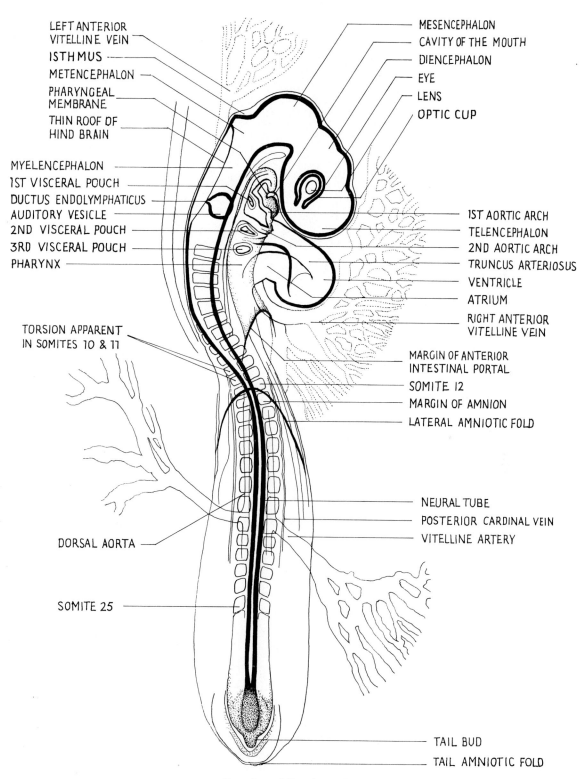

LEFT ANTERIOR VITELLINE VEIN
ISTHMUS
METENCEPHALON
PHARYNGEAL MEMBRANE
THIN ROOF OF HIND BRAIN

MYELENCEPHALON
1ST VISCERAL POUCH
DUCTUS ENDOLYMPHATICUS
AUDITORY VESICLE
2ND VISCERAL POUCH
3RD VISCERAL POUCH
PHARYNX

TORSION APPARENT IN SOMITES 10 & 11

DORSAL AORTA

SOMITE 25

MESENCEPHALON
CAVITY OF THE MOUTH
DIENCEPHALON
EYE
LENS
OPTIC CUP

1ST AORTIC ARCH
TELENCEPHALON
2ND AORTIC ARCH
TRUNCUS ARTERIOSUS
VENTRICLE
ATRIUM
RIGHT ANTERIOR VITELLINE VEIN

MARGIN OF ANTERIOR INTESTINAL PORTAL
SOMITE 12
MARGIN OF AMNION
LATERAL AMNIOTIC FOLD

NEURAL TUBE
POSTERIOR CARDINAL VEIN
VITELLINE ARTERY

TAIL BUD
TAIL AMNIOTIC FOLD

Drawing of Specimen 51

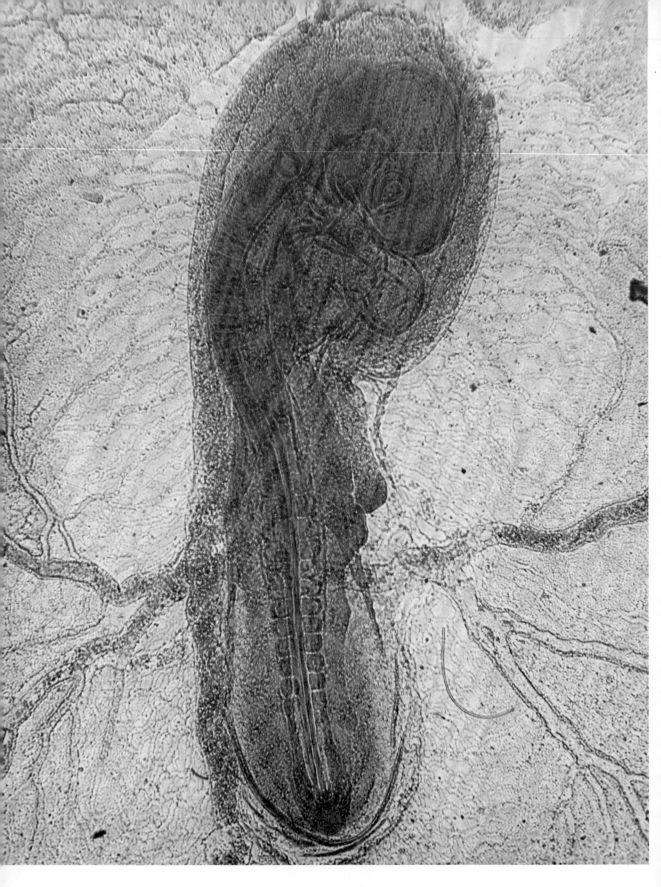

52. **Chick:** blastoderm, 30-somite, E. *mag. 25×*

61

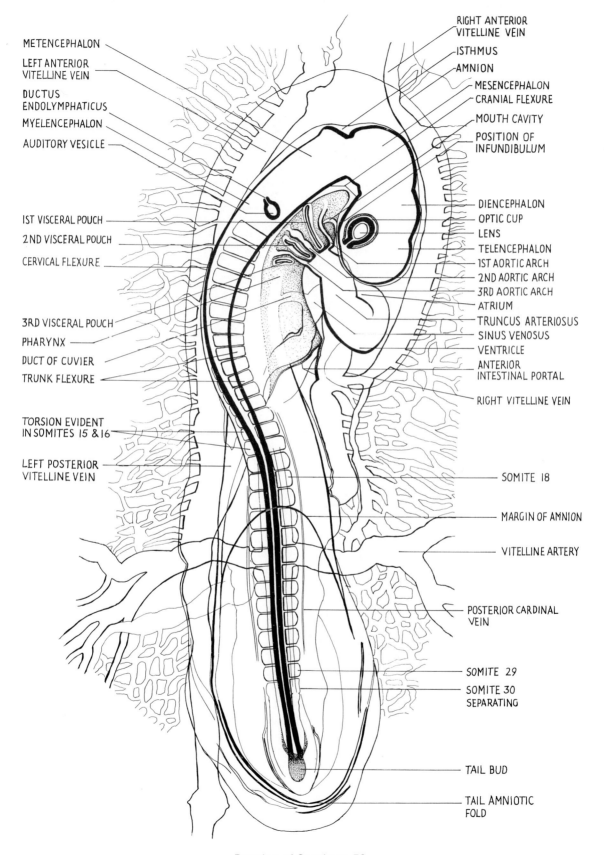

METENCEPHALON

LEFT ANTERIOR
VITELLINE VEIN

DUCTUS
ENDOLYMPHATICUS

MYELENCEPHALON

AUDITORY VESICLE

1ST VISCERAL POUCH

2ND VISCERAL POUCH

CERVICAL FLEXURE

3RD VISCERAL POUCH

PHARYNX

DUCT OF CUVIER

TRUNK FLEXURE

TORSION EVIDENT
IN SOMITES 15 & 16

LEFT POSTERIOR
VITELLINE VEIN

RIGHT ANTERIOR
VITELLINE VEIN

ISTHMUS

AMNION

MESENCEPHALON

CRANIAL FLEXURE

MOUTH CAVITY

POSITION OF
INFUNDIBULUM

DIENCEPHALON

OPTIC CUP

LENS

TELENCEPHALON

1ST AORTIC ARCH

2ND AORTIC ARCH

3RD AORTIC ARCH

ATRIUM

TRUNCUS ARTERIOSUS

SINUS VENOSUS

VENTRICLE

ANTERIOR
INTESTINAL PORTAL

RIGHT VITELLINE VEIN

SOMITE 18

MARGIN OF AMNION

VITELLINE ARTERY

POSTERIOR CARDINAL
VEIN

SOMITE 29

SOMITE 30
SEPARATING

TAIL BUD

TAIL AMNIOTIC
FOLD

Drawing of Specimen 52

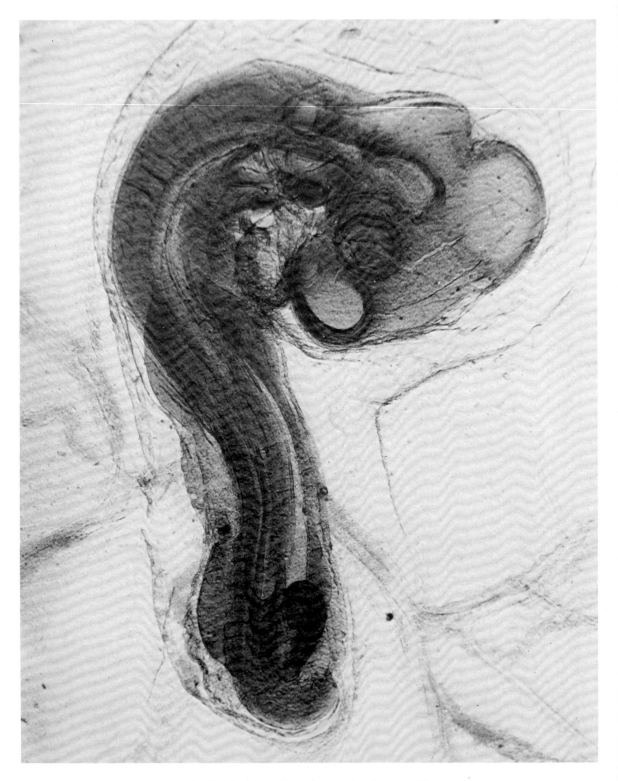

53. **Chick:** blastoderm, 35-somite, E. *mag. 30 ×*

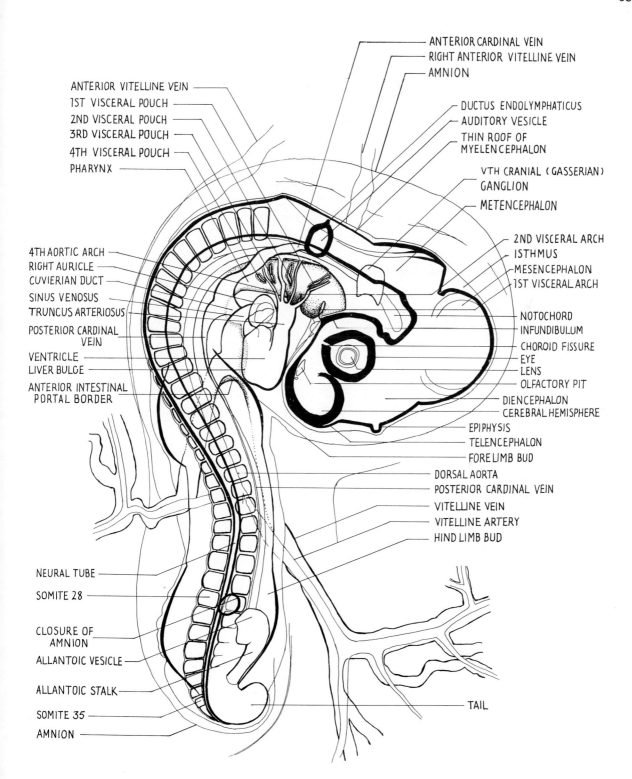

ANTERIOR CARDINAL VEIN
RIGHT ANTERIOR VITELLINE VEIN
AMNION

DUCTUS ENDOLYMPHATICUS
AUDITORY VESICLE
THIN ROOF OF
MYELENCEPHALON

VTH CRANIAL (GASSERIAN)
GANGLION
METENCEPHALON

2ND VISCERAL ARCH
ISTHMUS
MESENCEPHALON
1ST VISCERAL ARCH

NOTOCHORD
INFUNDIBULUM
CHOROID FISSURE
EYE
LENS
OLFACTORY PIT
DIENCEPHALON
CEREBRAL HEMISPHERE
EPIPHYSIS
TELENCEPHALON
FORE LIMB BUD

DORSAL AORTA
POSTERIOR CARDINAL VEIN
VITELLINE VEIN
VITELLINE ARTERY
HIND LIMB BUD

TAIL

ANTERIOR VITELLINE VEIN
1ST VISCERAL POUCH
2ND VISCERAL POUCH
3RD VISCERAL POUCH
4TH VISCERAL POUCH
PHARYNX

4TH AORTIC ARCH
RIGHT AURICLE
CUVIERIAN DUCT
SINUS VENOSUS
TRUNCUS ARTERIOSUS
POSTERIOR CARDINAL
VEIN
VENTRICLE
LIVER BULGE
ANTERIOR INTESTINAL
PORTAL BORDER

NEURAL TUBE
SOMITE 28
CLOSURE OF
AMNION
ALLANTOIC VESICLE
ALLANTOIC STALK
SOMITE 35
AMNION

Drawing of Specimen 53

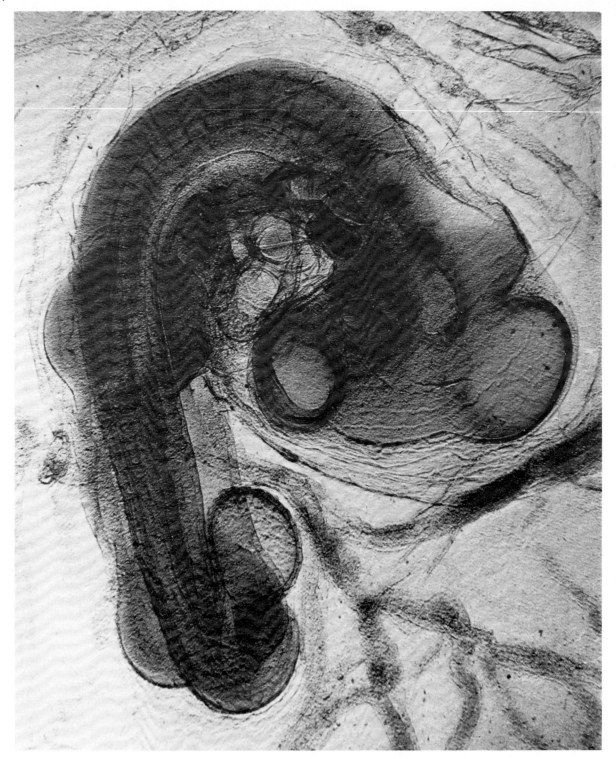

54. **Chick:** blastoderm, 40-somite, E. *mag. 30 ×*

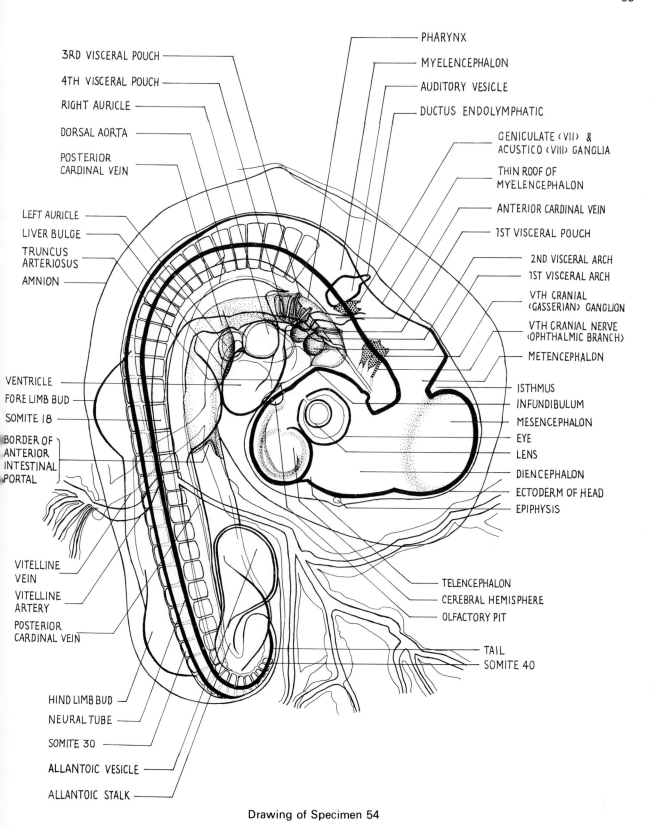

PHARYNX

MYELENCEPHALON

AUDITORY VESICLE

DUCTUS ENDOLYMPHATIC

GENICULATE ⟨VII⟩ &
ACUSTICO ⟨VIII⟩ GANGLIA

THIN ROOF OF
MYELENCEPHALON

ANTERIOR CARDINAL VEIN

1ST VISCERAL POUCH

2ND VISCERAL ARCH

1ST VISCERAL ARCH

VTH CRANIAL
⟨GASSERIAN⟩ GANGLION

VTH CRANIAL NERVE
⟨OPHTHALMIC BRANCH⟩

METENCEPHALON

ISTHMUS

INFUNDIBULUM

MESENCEPHALON

EYE

LENS

DIENCEPHALON

ECTODERM OF HEAD

EPIPHYSIS

TELENCEPHALON

CEREBRAL HEMISPHERE

OLFACTORY PIT

TAIL

SOMITE 40

3RD VISCERAL POUCH

4TH VISCERAL POUCH

RIGHT AURICLE

DORSAL AORTA

POSTERIOR
CARDINAL VEIN

LEFT AURICLE

LIVER BULGE

TRUNCUS
ARTERIOSUS

AMNION

VENTRICLE

FORE LIMB BUD

SOMITE 18

BORDER OF
ANTERIOR
INTESTINAL
PORTAL

VITELLINE
VEIN

VITELLINE
ARTERY

POSTERIOR
CARDINAL VEIN

HIND LIMB BUD

NEURAL TUBE

SOMITE 30

ALLANTOIC VESICLE

ALLANTOIC STALK

Drawing of Specimen 54

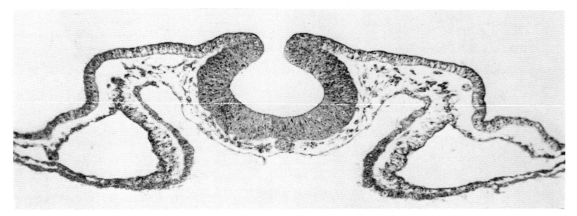

55. **Chick:** 6-somite stage, head region, T.S. *mag. 140 ×*

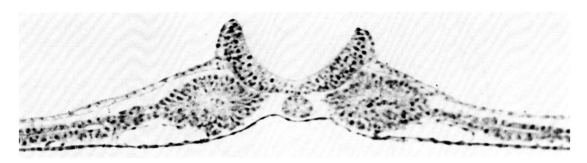

56. **Chick:** 6-somite stage, somitic region, T.S. *mag. 200 ×*

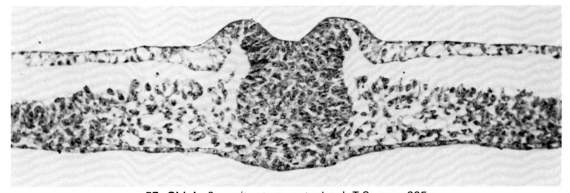

57. **Chick:** 6-somite stage, notochord, T.S. *mag. 225 ×*

58. **Chick:** 6-somite stage, primitive streak, T.S. *mag. 200 ×*

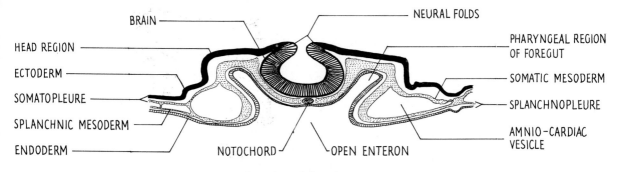

BRAIN — NEURAL FOLDS

HEAD REGION — PHARYNGEAL REGION OF FOREGUT

ECTODERM — SOMATIC MESODERM

SOMATOPLEURE — SPLANCHNOPLEURE

SPLANCHNIC MESODERM — AMNIO–CARDIAC VESICLE

ENDODERM — NOTOCHORD — OPEN ENTERON

Drawing of Specimen 55

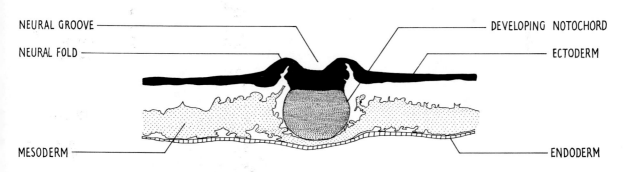

NEURAL GROOVE — SOMITE

NEURAL FOLD — MESENCHYME FORMING DORSAL AORTA

CENTRAL CORE OF CELLS — SOMATIC MESODERM

ECTODERM — SOMATOPLEURE

COELOM — SPLANCHNOPLEURE

SPLANCHNIC MESODERM — NOTOCHORD — OPEN ENTERON

ENDODERM

Drawing of Specimen 56

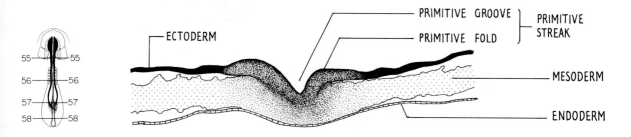

NEURAL GROOVE — DEVELOPING NOTOCHORD

NEURAL FOLD — ECTODERM

MESODERM — ENDODERM

Drawing of Specimen 57

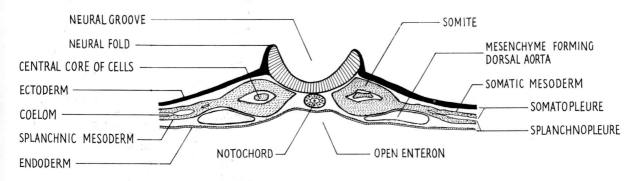

ECTODERM — PRIMITIVE GROOVE ⎤ PRIMITIVE STREAK

PRIMITIVE FOLD ⎦

MESODERM

ENDODERM

55 — 55
56 — 56
57 — 57
58 — 58

Drawing of Specimen 58

68

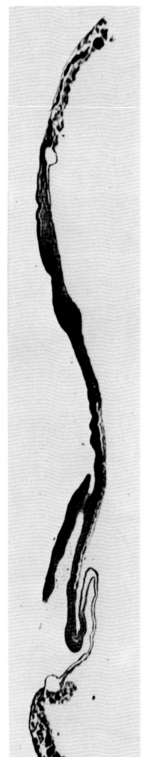

59. **Chick:** 6-somite stage, U.L.S. *mag. 38* ×

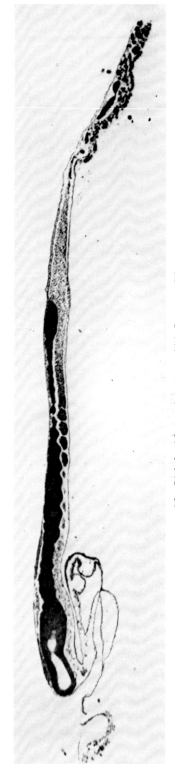

60. **Chick:** 10-somite stage, U.L.S. *mag. 28* ×

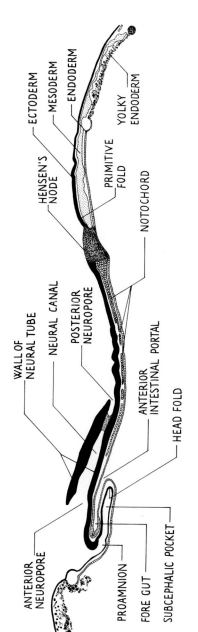

ANTERIOR
NEUROPORE

WALL OF
NEURAL TUBE

NEURAL CANAL

POSTERIOR
NEUROPORE

HENSEN'S
NODE

ECTODERM

MESODERM

ENDODERM

YOLKY
ENDODERM

PRIMITIVE
FOLD

NOTOCHORD

ANTERIOR
INTESTINAL PORTAL

HEAD FOLD

PROAMNION

FORE GUT

SUBCEPHALIC POCKET

Drawing of Specimen 59

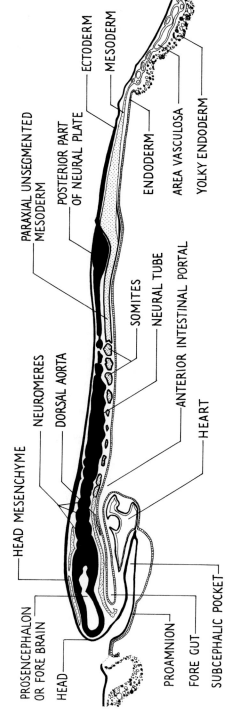

HEAD MESENCHYME

NEUROMERES

DORSAL AORTA

PARAXIAL UNSEGMENTED
MESODERM

POSTERIOR PART
OF NEURAL PLATE

ECTODERM

MESODERM

ENDODERM

AREA VASCULOSA

YOLKY ENDODERM

SOMITES

NEURAL TUBE

ANTERIOR INTESTINAL PORTAL

HEART

PROSENCEPHALON
OR FORE BRAIN

HEAD

PROAMNION

FORE GUT

SUBCEPHALIC POCKET

Drawing of Specimen 60

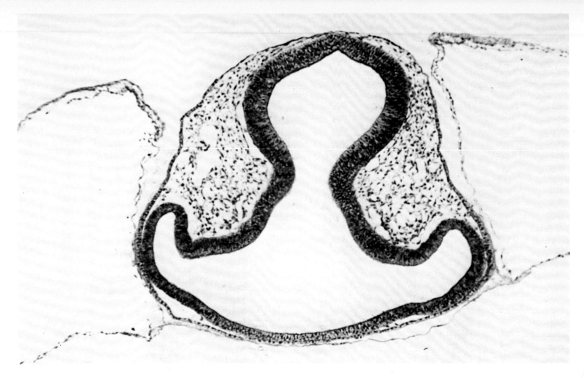

61. **Chick:** 10-somite stage, forebrain region, T.S. *mag. 100×*

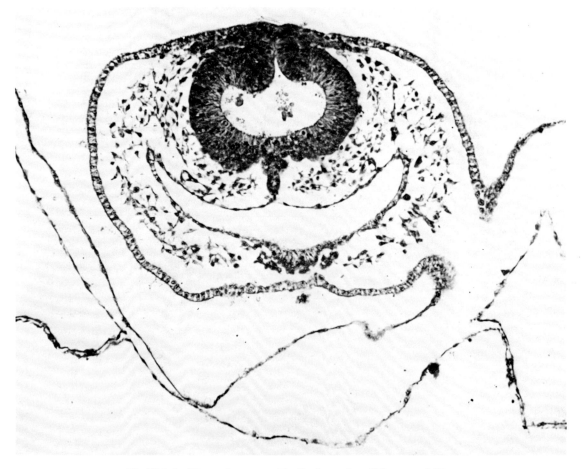

62. **Chick:** 10-somite stage, hindbrain region, T.S. *mag. 200×*

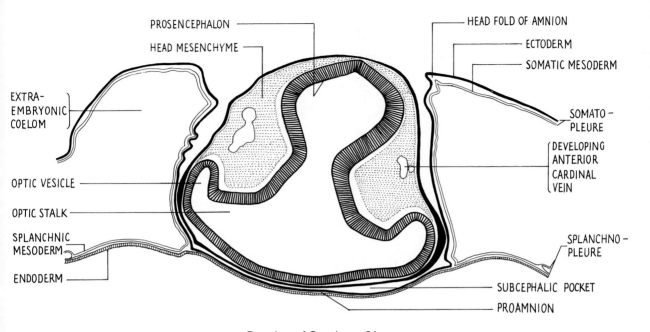

PROSENCEPHALON

HEAD MESENCHYME

HEAD FOLD OF AMNION

ECTODERM

SOMATIC MESODERM

EXTRA-EMBRYONIC COELOM

SOMATO-PLEURE

DEVELOPING ANTERIOR CARDINAL VEIN

OPTIC VESICLE

OPTIC STALK

SPLANCHNIC MESODERM

ENDODERM

SPLANCHNO-PLEURE

SUBCEPHALIC POCKET

PROAMNION

Drawing of Specimen 61

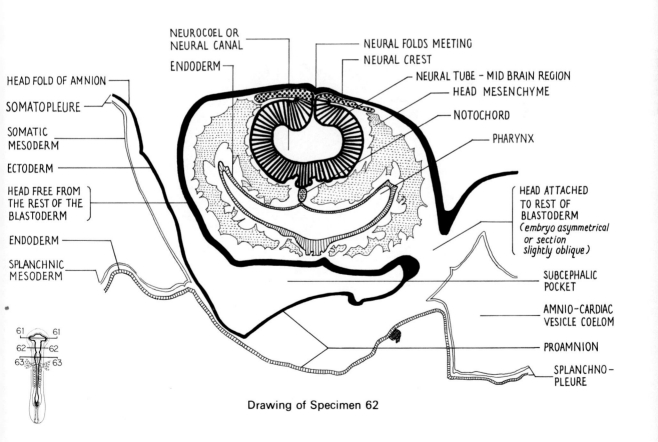

NEUROCOEL OR NEURAL CANAL

ENDODERM

NEURAL FOLDS MEETING

NEURAL CREST

NEURAL TUBE – MID BRAIN REGION

HEAD MESENCHYME

NOTOCHORD

PHARYNX

HEAD FOLD OF AMNION

SOMATOPLEURE

SOMATIC MESODERM

ECTODERM

HEAD FREE FROM THE REST OF THE BLASTODERM

ENDODERM

SPLANCHNIC MESODERM

HEAD ATTACHED TO REST OF BLASTODERM (embryo asymmetrical or section slightly oblique)

SUBCEPHALIC POCKET

AMNIO-CARDIAC VESICLE COELOM

PROAMNION

SPLANCHNO-PLEURE

61   61
62   62
63   63

Drawing of Specimen 62

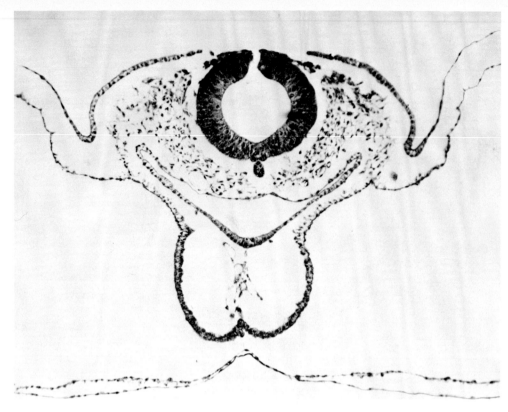

63. **Chick:** 10-somite stage, heart region, T.S. *mag. 150 ×*

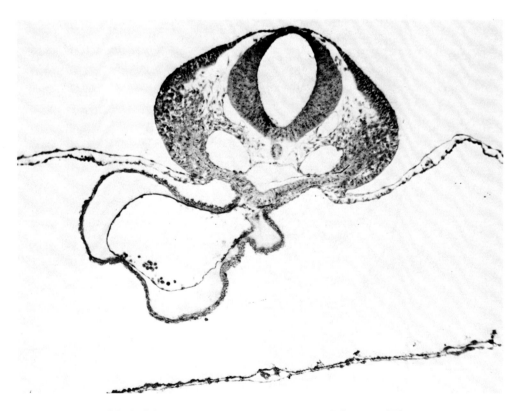

64. **Chick:** 13-somite stage, heart region, T.S. *mag. 150 ×*

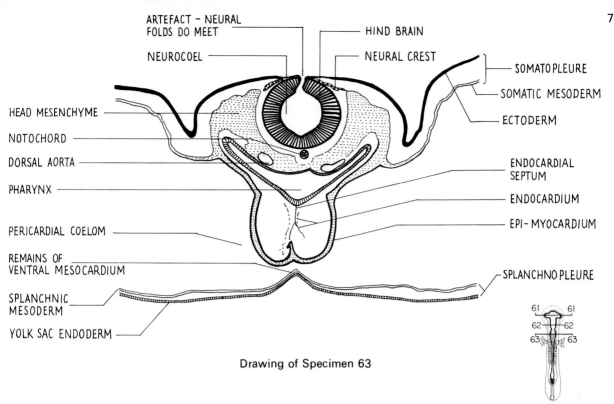

ARTEFACT - NEURAL FOLDS DO MEET

HIND BRAIN

NEUROCOEL

NEURAL CREST

SOMATOPLEURE

SOMATIC MESODERM

ECTODERM

HEAD MESENCHYME

NOTOCHORD

DORSAL AORTA

PHARYNX

ENDOCARDIAL SEPTUM

ENDOCARDIUM

EPI-MYOCARDIUM

PERICARDIAL COELOM

REMAINS OF VENTRAL MESOCARDIUM

SPLANCHNOPLEURE

SPLANCHNIC MESODERM

YOLK SAC ENDODERM

Drawing of Specimen 63

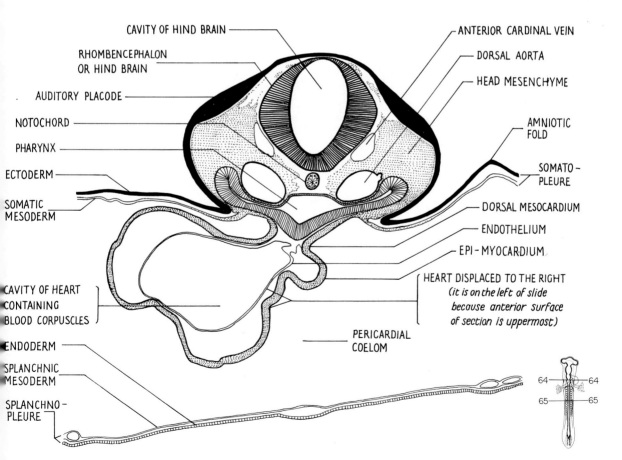

CAVITY OF HIND BRAIN

ANTERIOR CARDINAL VEIN

RHOMBENCEPHALON OR HIND BRAIN

DORSAL AORTA

HEAD MESENCHYME

AUDITORY PLACODE

NOTOCHORD

AMNIOTIC FOLD

PHARYNX

ECTODERM

SOMATO-PLEURE

SOMATIC MESODERM

DORSAL MESOCARDIUM

ENDOTHELIUM

EPI-MYOCARDIUM

CAVITY OF HEART CONTAINING BLOOD CORPUSCLES

HEART DISPLACED TO THE RIGHT
(it is on the left of slide because anterior surface of section is uppermost)

ENDODERM

SPLANCHNIC MESODERM

PERICARDIAL COELOM

SPLANCHNO-PLEURE

Drawing of Specimen 64

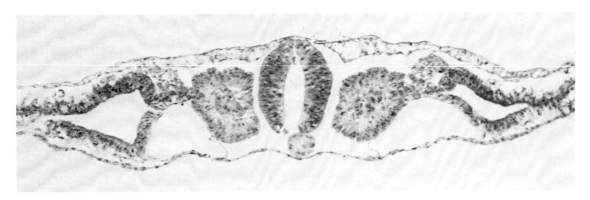

65. **Chick:** 13-somite stage, posterior trunk region, T.S. *mag. 175 ×*

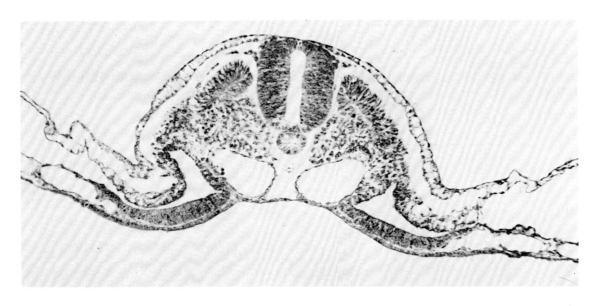

66. **Chick:** 17-somite stage, trunk region, T.S. *mag. 150 ×*

75

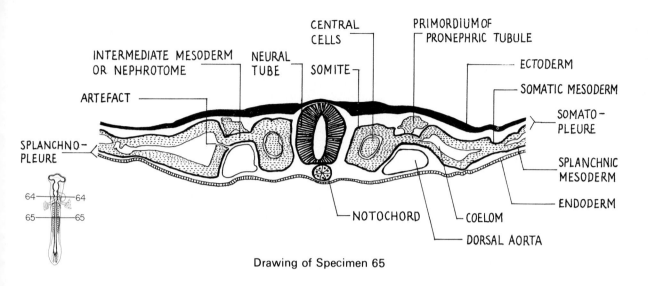

CENTRAL CELLS

PRIMORDIUM OF PRONEPHRIC TUBULE

INTERMEDIATE MESODERM OR NEPHROTOME

NEURAL TUBE

SOMITE

ECTODERM

SOMATIC MESODERM

SOMATO-PLEURE

ARTEFACT

SPLANCHNO-PLEURE

SPLANCHNIC MESODERM

ENDODERM

NOTOCHORD

COELOM

DORSAL AORTA

Drawing of Specimen 65

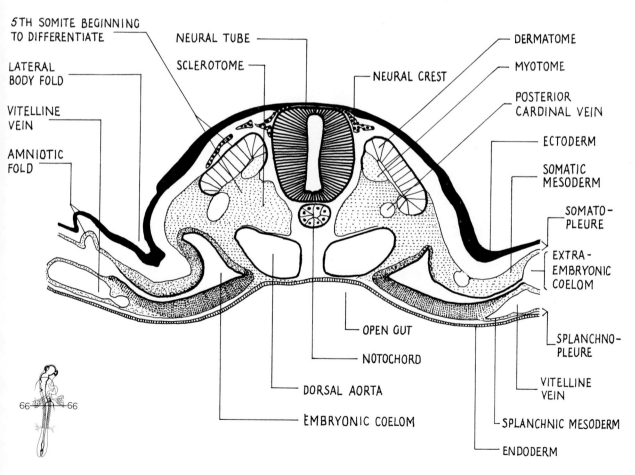

5TH SOMITE BEGINNING TO DIFFERENTIATE

NEURAL TUBE

SCLEROTOME

DERMATOME

NEURAL CREST

MYOTOME

LATERAL BODY FOLD

POSTERIOR CARDINAL VEIN

VITELLINE VEIN

ECTODERM

AMNIOTIC FOLD

SOMATIC MESODERM

SOMATO-PLEURE

EXTRA-EMBRYONIC COELOM

OPEN GUT

NOTOCHORD

SPLANCHNO-PLEURE

VITELLINE VEIN

DORSAL AORTA

EMBRYONIC COELOM

SPLANCHNIC MESODERM

ENDODERM

Drawing of Specimen 66

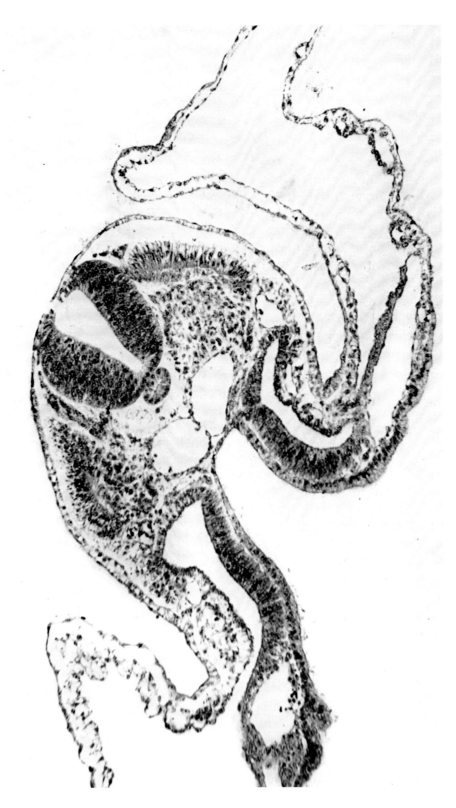

67. **Chick**: 21-somite stage, trunk region, T.S. *mag. 200* ×

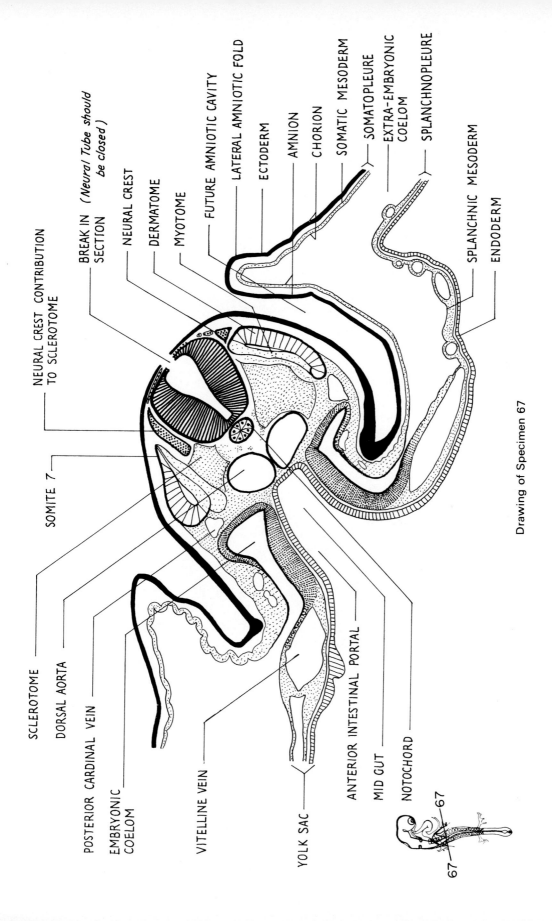

SCLEROTOME

DORSAL AORTA

POSTERIOR CARDINAL VEIN

EMBRYONIC COELOM

VITELLINE VEIN

YOLK SAC

ANTERIOR INTESTINAL PORTAL

MID GUT

NOTOCHORD

NEURAL CREST CONTRIBUTION TO SCLEROTOME

SOMITE 7

BREAK IN (Neural Tube should SECTION be closed)

NEURAL CREST

DERMATOME

MYOTOME

FUTURE AMNIOTIC CAVITY

LATERAL AMNIOTIC FOLD

ECTODERM

AMNION

CHORION

SOMATIC MESODERM

SOMATOPLEURE

EXTRA-EMBRYONIC COELOM

SPLANCHNOPLEURE

SPLANCHNIC MESODERM

ENDODERM

Drawing of Specimen 67

67

67

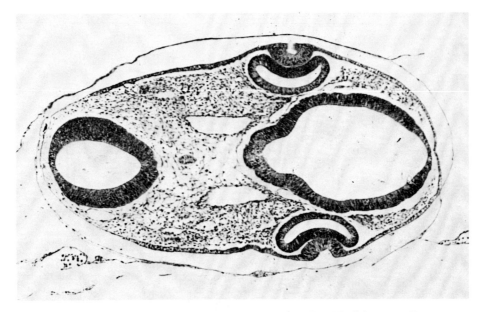

68. **Chick:** 24-somite stage, fore and hind brain, T.S. (1) *mag. 45 ×*

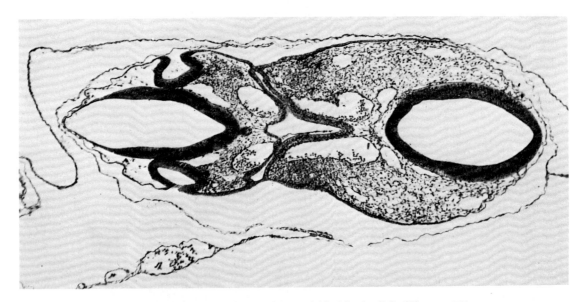

69. **Chick:** 24-somite stage, fore and hind brain, T.S. (2) *mag. 70 ×*

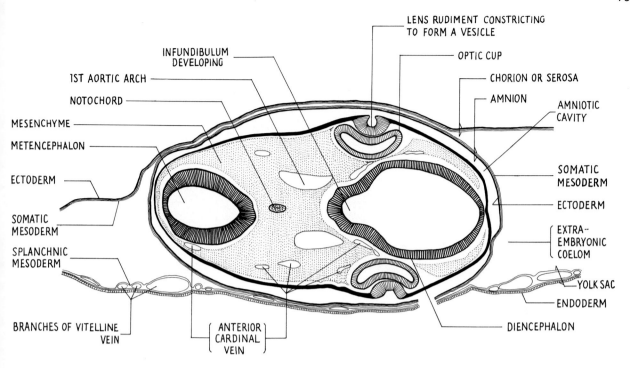

LENS RUDIMENT CONSTRICTING TO FORM A VESICLE

OPTIC CUP

CHORION OR SEROSA

AMNION

AMNIOTIC CAVITY

INFUNDIBULUM DEVELOPING

1ST AORTIC ARCH

NOTOCHORD

MESENCHYME

METENCEPHALON

ECTODERM

SOMATIC MESODERM

SPLANCHNIC MESODERM

BRANCHES OF VITELLINE VEIN

ANTERIOR CARDINAL VEIN

SOMATIC MESODERM

ECTODERM

EXTRA-EMBRYONIC COELOM

YOLK SAC

ENDODERM

DIENCEPHALON

Drawing of Specimen 68

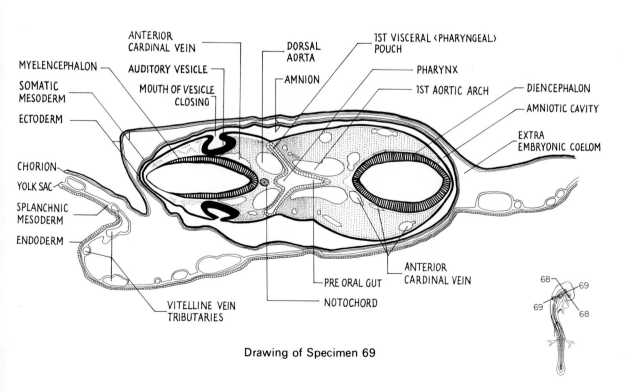

ANTERIOR CARDINAL VEIN

DORSAL AORTA

1ST VISCERAL ‹PHARYNGEAL› POUCH

MYELENCEPHALON

AUDITORY VESICLE

AMNION

PHARYNX

DIENCEPHALON

SOMATIC MESODERM

MOUTH OF VESICLE CLOSING

1ST AORTIC ARCH

AMNIOTIC CAVITY

ECTODERM

EXTRA EMBRYONIC COELOM

CHORION

YOLK SAC

SPLANCHNIC MESODERM

ENDODERM

ANTERIOR CARDINAL VEIN

VITELLINE VEIN TRIBUTARIES

PRE ORAL GUT

NOTOCHORD

68 69

69 68

Drawing of Specimen 69

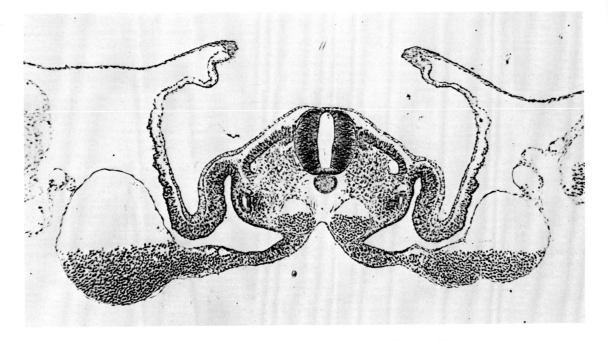

70. **Chick:** 27-somite stage, trunk region, T.S. *mag. 80×*

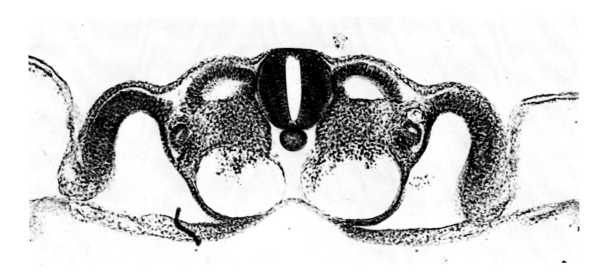

71. **Chick:** 27-somite stage, posterior trunk region, T.S. mag. 95×

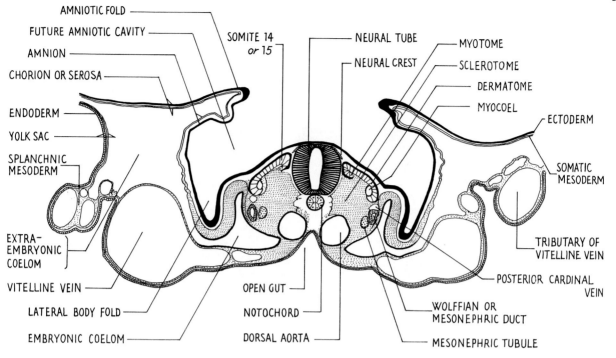

AMNIOTIC FOLD

FUTURE AMNIOTIC CAVITY

AMNION

CHORION OR SEROSA

ENDODERM

YOLK SAC

SPLANCHNIC MESODERM

EXTRA-EMBRYONIC COELOM

VITELLINE VEIN

LATERAL BODY FOLD

EMBRYONIC COELOM

SOMITE 14 or 15

OPEN GUT

NOTOCHORD

DORSAL AORTA

NEURAL TUBE

NEURAL CREST

MYOTOME

SCLEROTOME

DERMATOME

MYOCOEL

ECTODERM

SOMATIC MESODERM

TRIBUTARY OF VITELLINE VEIN

POSTERIOR CARDINAL VEIN

WOLFFIAN OR MESONEPHRIC DUCT

MESONEPHRIC TUBULE

Drawing of Specimen 70

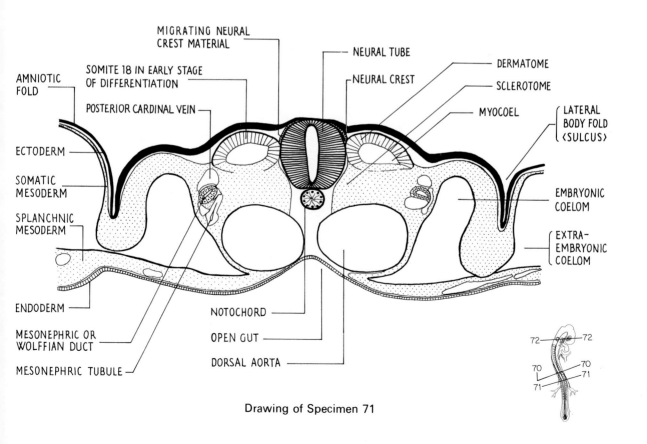

MIGRATING NEURAL CREST MATERIAL

AMNIOTIC FOLD

SOMITE 18 IN EARLY STAGE OF DIFFERENTIATION

POSTERIOR CARDINAL VEIN

ECTODERM

SOMATIC MESODERM

SPLANCHNIC MESODERM

ENDODERM

MESONEPHRIC OR WOLFFIAN DUCT

MESONEPHRIC TUBULE

NEURAL TUBE

NEURAL CREST

DERMATOME

SCLEROTOME

MYOCOEL

LATERAL BODY FOLD ⟨SULCUS⟩

EMBRYONIC COELOM

EXTRA-EMBRYONIC COELOM

NOTOCHORD

OPEN GUT

DORSAL AORTA

Drawing of Specimen 71

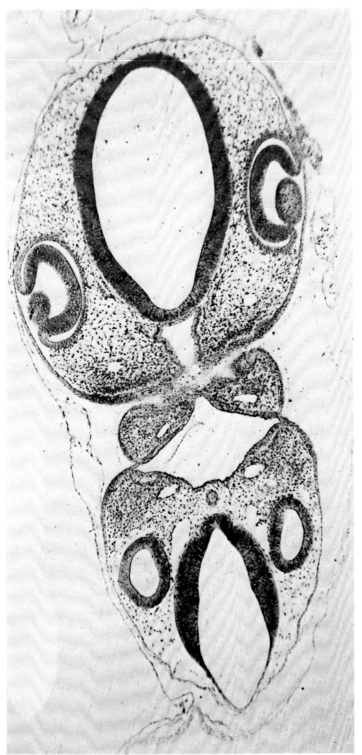

72. **Chick**: 27-stomite stage, eye and ear region, T.S. *mag. 90* ×

83

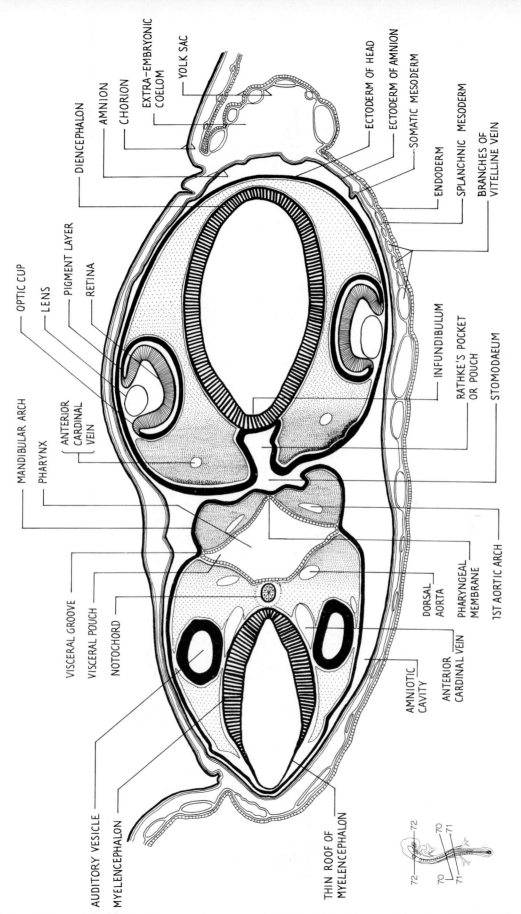

DIENCEPHALON

AMNION

CHORION

EXTRA-EMBRYONIC COELOM

YOLK SAC

ECTODERM OF HEAD

ECTODERM OF AMNION

SOMATIC MESODERM

ENDODERM

SPLANCHNIC MESODERM

BRANCHES OF VITELLINE VEIN

OPTIC CUP

LENS

PIGMENT LAYER

RETINA

MANDIBULAR ARCH

PHARYNX

ANTERIOR CARDINAL VEIN

INFUNDIBULUM

RATHKE'S POCKET OR POUCH

STOMODAEUM

DORSAL AORTA

PHARYNGEAL MEMBRANE

1ST AORTIC ARCH

AMNIOTIC CAVITY

ANTERIOR CARDINAL VEIN

VISCERAL GROOVE

VISCERAL POUCH

NOTOCHORD

AUDITORY VESICLE

MYELENCEPHALON

THIN ROOF OF MYELENCEPHALON

Drawing of Specimen 72

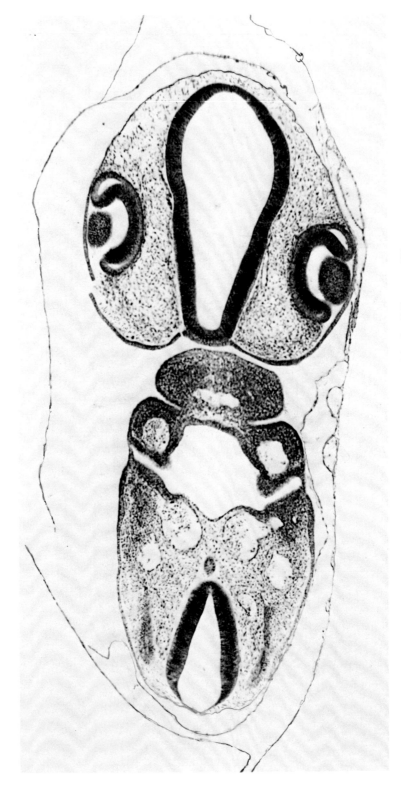

73. **Chick**: 30-somite stage, fore and hind brain, T.S. *mag. 75* ×

85

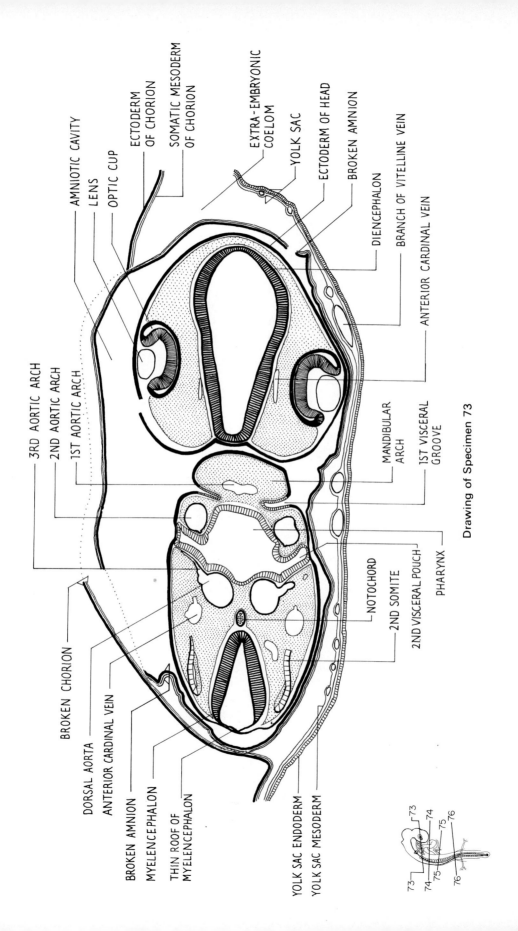

Drawing of Specimen 73

86

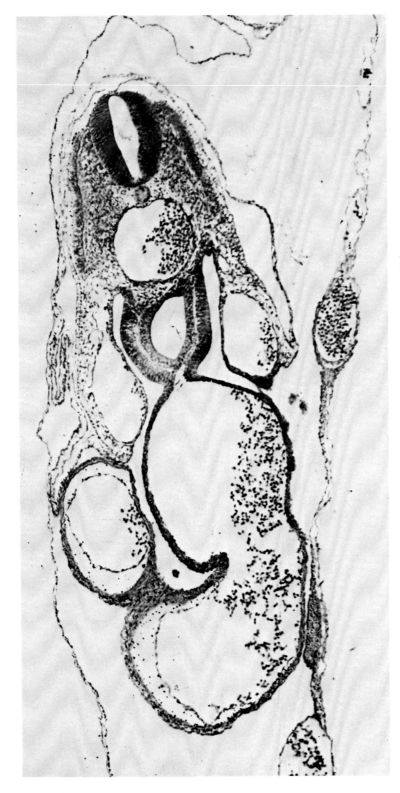

74. **Chick**: 30-somite stage, heart region, T.S. *mag. 130 ×*

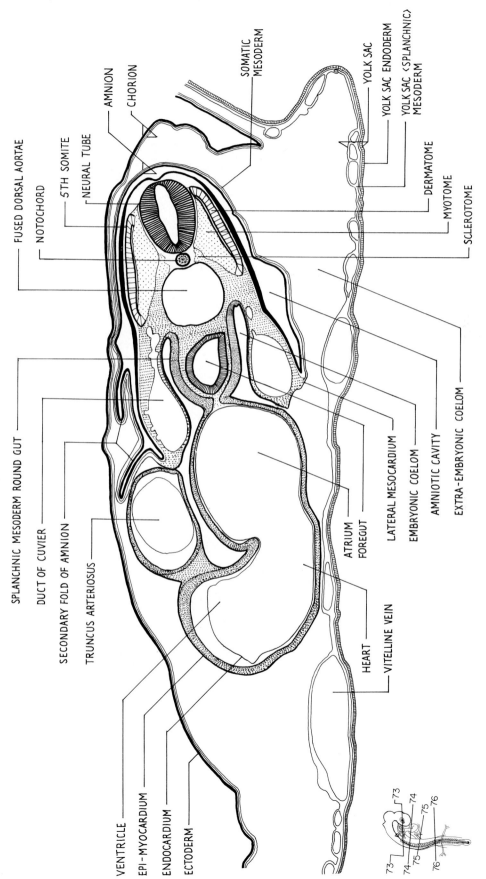

SOMATIC MESODERM

YOLK SAC

YOLK SAC ENDODERM

YOLK SAC (SPLANCHNIC) MESODERM

AMNION

CHORION

DERMATOME

5TH SOMITE

NEURAL TUBE

MYOTOME

FUSED DORSAL AORTAE

NOTOCHORD

SCLEROTOME

SPLANCHNIC MESODERM ROUND GUT

DUCT OF CUVIER

SECONDARY FOLD OF AMNION

TRUNCUS ARTERIOSUS

ATRIUM

FOREGUT

LATERAL MESOCARDIUM

EMBRYONIC COELOM

AMNIOTIC CAVITY

EXTRA-EMBRYONIC COELOM

VENTRICLE

EPI-MYOCARDIUM

ENDOCARDIUM

ECTODERM

HEART

VITELLINE VEIN

Drawing of Specimen 74

73
74
75
76

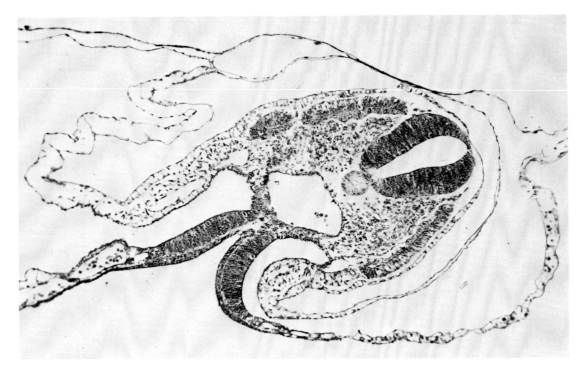

75. **Chick:** 30-somite stage, anterior trunk region, T.S. *mag. 125 ×*

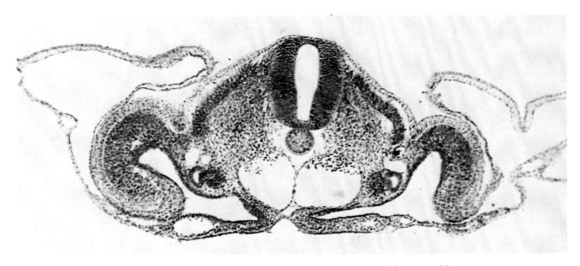

76. **Chick:** 30-somite stage, posterior trunk region, T.S. *mag. 85 ×*

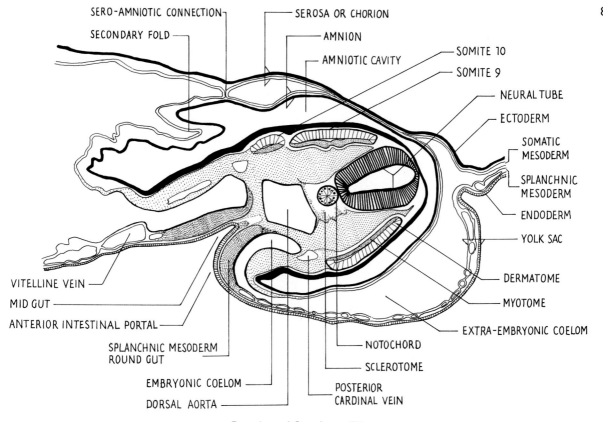

SERO-AMNIOTIC CONNECTION

SECONDARY FOLD

SEROSA OR CHORION

AMNION

AMNIOTIC CAVITY

SOMITE 10

SOMITE 9

NEURAL TUBE

ECTODERM

SOMATIC MESODERM

SPLANCHNIC MESODERM

ENDODERM

YOLK SAC

DERMATOME

MYOTOME

EXTRA-EMBRYONIC COELOM

VITELLINE VEIN

MID GUT

ANTERIOR INTESTINAL PORTAL

SPLANCHNIC MESODERM ROUND GUT

EMBRYONIC COELOM

DORSAL AORTA

NOTOCHORD

SCLEROTOME

POSTERIOR CARDINAL VEIN

Drawing of Specimen 75

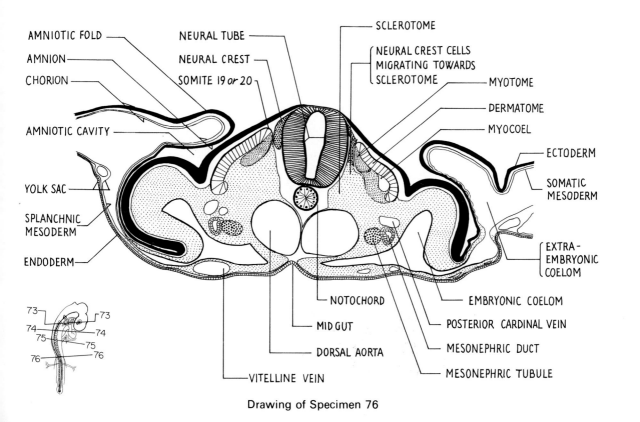

AMNIOTIC FOLD

AMNION

CHORION

AMNIOTIC CAVITY

YOLK SAC

SPLANCHNIC MESODERM

ENDODERM

NEURAL TUBE

NEURAL CREST

SOMITE 19 or 20

SCLEROTOME

NEURAL CREST CELLS MIGRATING TOWARDS SCLEROTOME

MYOTOME

DERMATOME

MYOCOEL

ECTODERM

SOMATIC MESODERM

EXTRA-EMBRYONIC COELOM

NOTOCHORD

MID GUT

DORSAL AORTA

VITELLINE VEIN

EMBRYONIC COELOM

POSTERIOR CARDINAL VEIN

MESONEPHRIC DUCT

MESONEPHRIC TUBULE

Drawing of Specimen 76

90

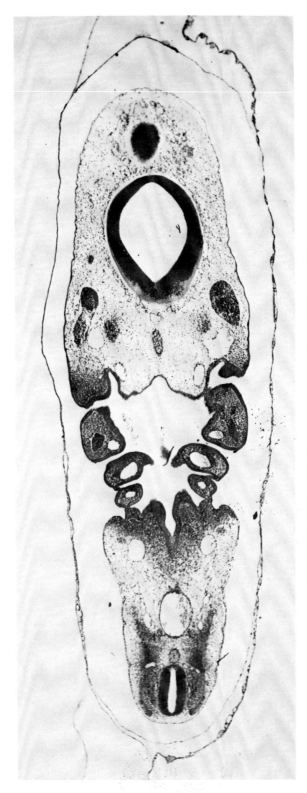

77. **Chick**: 36-somite stage, pharyngeal region, T.S. *mag. 40 ×*

91

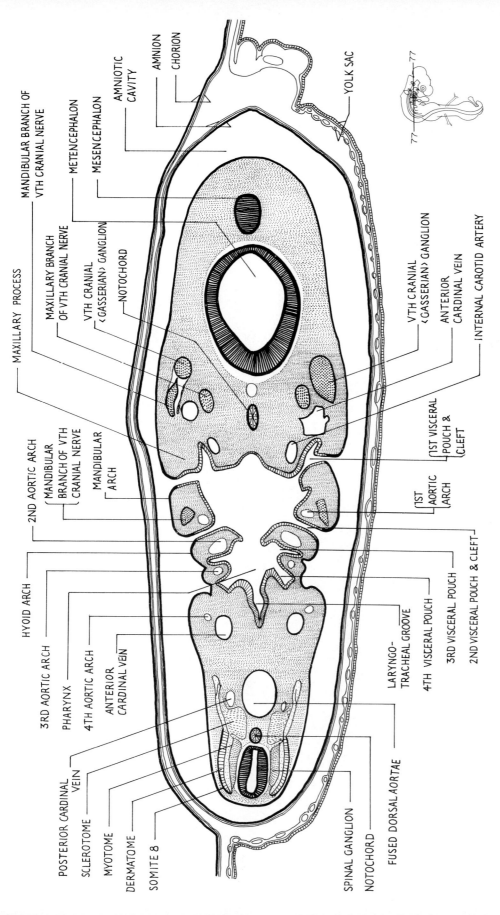

AMNION

CHORION

YOLK SAC

MANDIBULAR BRANCH OF
VTH CRANIAL NERVE

METENCEPHALON

MESENCEPHALON

AMNIOTIC
CAVITY

MAXILLARY BRANCH
OF VTH CRANIAL NERVE

VTH CRANIAL
〈GASSERIAN〉 GANGLION

NOTOCHORD

MAXILLARY PROCESS

VTH CRANIAL
〈GASSERIAN〉 GANGLION

ANTERIOR
CARDINAL VEIN

INTERNAL CAROTID ARTERY

2ND AORTIC ARCH

MANDIBULAR
BRANCH OF VTH
CRANIAL NERVE

MANDIBULAR
ARCH

1ST VISCERAL
POUCH &
CLEFT

1ST
AORTIC
ARCH

HYOID ARCH

3RD AORTIC ARCH

PHARYNX

4TH AORTIC ARCH

ANTERIOR
CARDINAL VEIN

LARYNGO-
TRACHEAL GROOVE

4TH VISCERAL POUCH

3RD VISCERAL POUCH

2ND VISCERAL POUCH & CLEFT

POSTERIOR CARDINAL VEIN

SCLEROTOME

MYOTOME

DERMATOME

SOMITE 8

SPINAL GANGLION

NOTOCHORD

FUSED DORSAL AORTAE

Drawing of Specimen 77

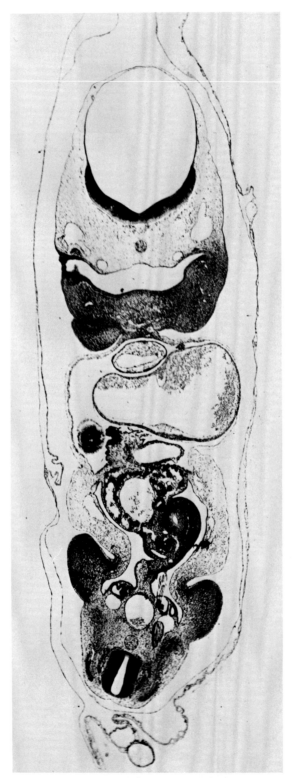

78. **Chick:** 36-somite stage, hind-brain region, T.S. *mag. 40* ×

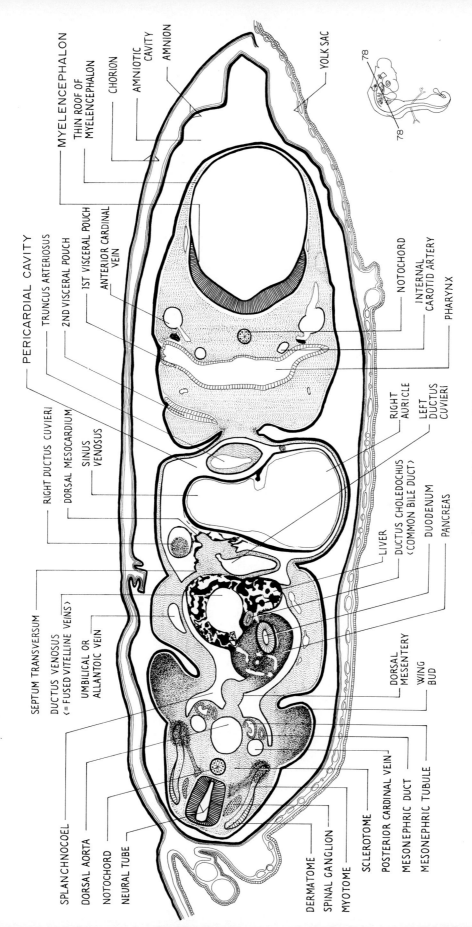

SPLANCHNOCOEL

DORSAL AORTA

NOTOCHORD

NEURAL TUBE

DERMATOME

SPINAL GANGLION

MYOTOME

SCLEROTOME

POSTERIOR CARDINAL VEIN

MESONEPHRIC DUCT

MESONEPHRIC TUBULE

SEPTUM TRANSVERSUM

DUCTUS VENOSUS
(= FUSED VITELLINE VEINS)

UMBILICAL OR
ALLANTOIC VEIN

RIGHT DUCTUS CUVIERI

DORSAL MESOCARDIUM

SINUS
VENOSUS

PERICARDIAL CAVITY

TRUNCUS ARTERIOSUS

2ND VISCERAL POUCH

1ST VISCERAL POUCH

ANTERIOR CARDINAL
VEIN

MYELENCEPHALON

THIN ROOF OF
MYELENCEPHALON

CHORION

AMNIOTIC
CAVITY

AMNION

YOLK SAC

NOTOCHORD

INTERNAL
CAROTID ARTERY

PHARYNX

RIGHT
AURICLE

LEFT
DUCTUS
CUVIERI

LIVER

DUCTUS CHOLEDOCHUS
(COMMON BILE DUCT)

DUODENUM

PANCREAS

DORSAL
MESENTERY

WING
BUD

Drawing of Specimen 78

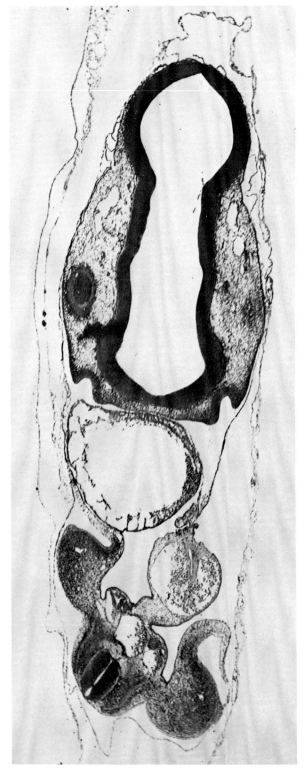

79. **Chick**: 36-somite stage, olfactory pit region, T.S. *mag. 45* ×

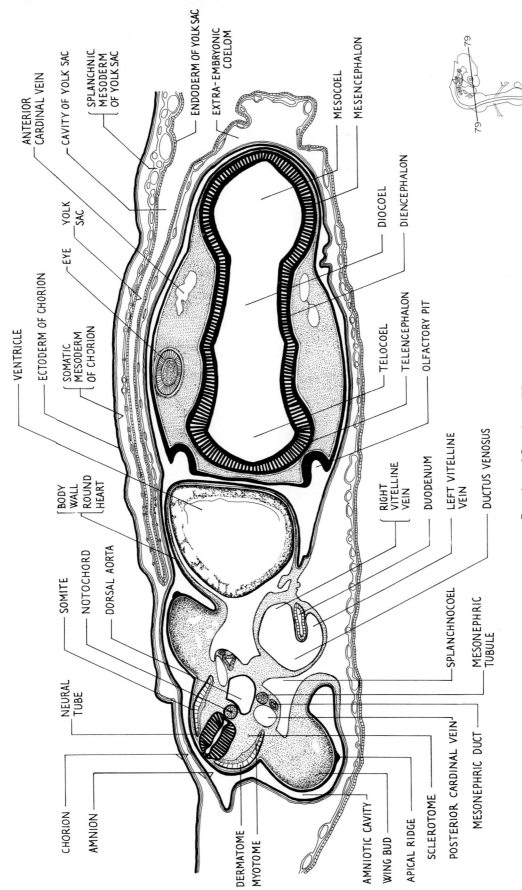

ANTERIOR CARDINAL VEIN

CAVITY OF YOLK SAC

SPLANCHNIC MESODERM OF YOLK SAC

ENDODERM OF YOLK SAC

EXTRA-EMBRYONIC COELOM

MESOCOEL

MESENCEPHALON

VENTRICLE

ECTODERM OF CHORION

SOMATIC MESODERM OF CHORION

EYE

YOLK SAC

DIOCOEL

DIENCEPHALON

TELOCOEL

TELENCEPHALON

OLFACTORY PIT

BODY WALL

ROUND HEART

RIGHT VITELLINE VEIN

DUODENUM

LEFT VITELLINE VEIN

DUCTUS VENOSUS

SOMITE

NOTOCHORD

DORSAL AORTA

SPLANCHNOCOEL

MESONEPHRIC TUBULE

NEURAL TUBE

CHORION

AMNION

DERMATOME

MYOTOME

AMNIOTIC CAVITY

WING BUD

APICAL RIDGE

SCLEROTOME

POSTERIOR CARDINAL VEIN

MESONEPHRIC DUCT

Drawing of Specimen 79

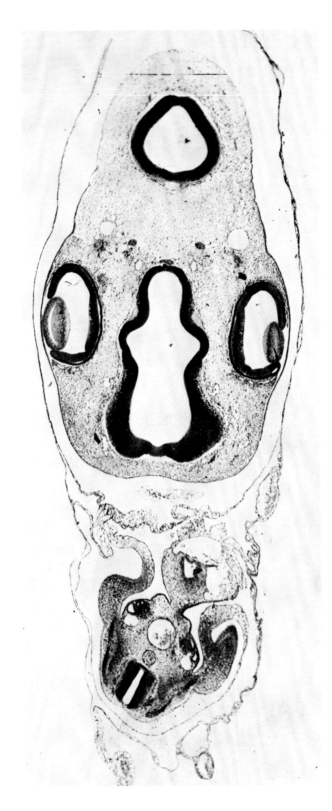

80. **Chick**: 36-somite stage, optic region, T.S. *mag. 40* ×

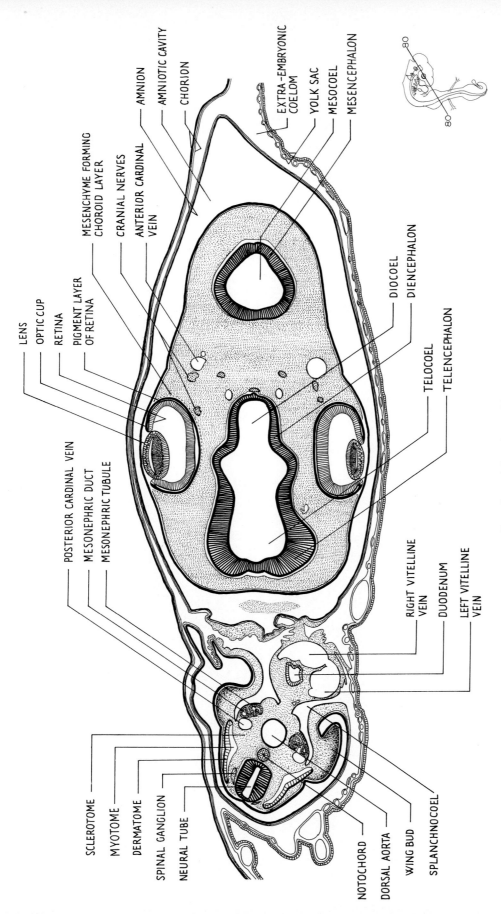

LENS

OPTIC CUP

RETINA

PIGMENT LAYER
OF RETINA

MESENCHYME FORMING
CHOROID LAYER

CRANIAL NERVES

ANTERIOR CARDINAL
VEIN

AMNION

AMNIOTIC CAVITY

CHORION

EXTRA-EMBRYONIC
COELOM

YOLK SAC

MESOCOEL

MESENCEPHALON

DIOCOEL

DIENCEPHALON

TELOCOEL

TELENCEPHALON

POSTERIOR CARDINAL VEIN

MESONEPHRIC DUCT

MESONEPHRIC TUBULE

RIGHT VITELLINE
VEIN

DUODENUM

LEFT VITELLINE
VEIN

SCLEROTOME

MYOTOME

DERMATOME

SPINAL GANGLION

NEURAL TUBE

NOTOCHORD

DORSAL AORTA

WING BUD

SPLANCHNOCOEL

Drawing of Specimen 80

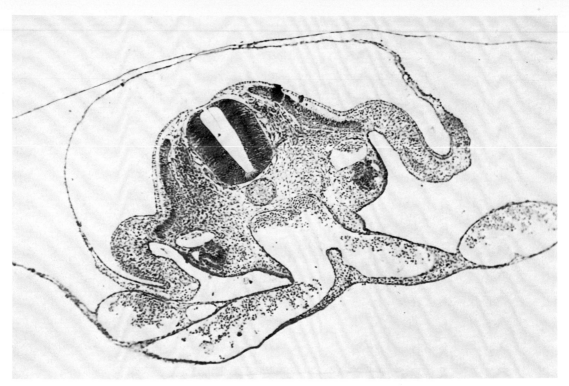

81. **Chick:** 36-somite stage, trunk region, T.S. *mag. 75×*

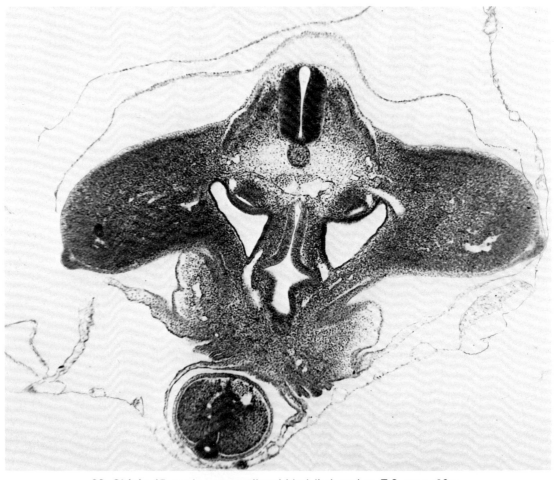

82. **Chick:** 45-somite stage, tail and hind-limb region, T.S. *mag. 60×*

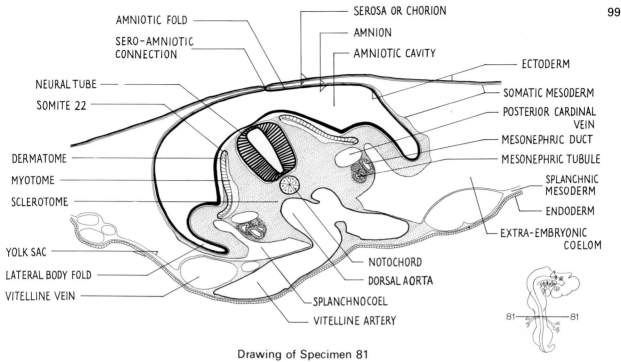

AMNIOTIC FOLD
SERO–AMNIOTIC CONNECTION
SEROSA OR CHORION
AMNION
AMNIOTIC CAVITY
ECTODERM
NEURAL TUBE
SOMITE 22
SOMATIC MESODERM
POSTERIOR CARDINAL VEIN
MESONEPHRIC DUCT
MESONEPHRIC TUBULE
DERMATOME
MYOTOME
SCLEROTOME
SPLANCHNIC MESODERM
ENDODERM
EXTRA-EMBRYONIC COELOM
YOLK SAC
LATERAL BODY FOLD
VITELLINE VEIN
NOTOCHORD
DORSAL AORTA
SPLANCHNOCOEL
VITELLINE ARTERY

81 — 81

Drawing of Specimen 81

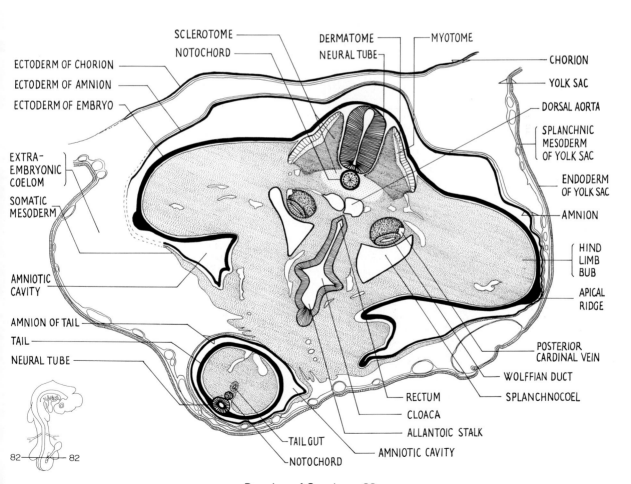

SCLEROTOME
NOTOCHORD
DERMATOME
NEURAL TUBE
MYOTOME
CHORION
ECTODERM OF CHORION
ECTODERM OF AMNION
ECTODERM OF EMBRYO
YOLK SAC
DORSAL AORTA
SPLANCHNIC MESODERM OF YOLK SAC
ENDODERM OF YOLK SAC
EXTRA-EMBRYONIC COELOM
SOMATIC MESODERM
AMNION
HIND LIMB BUB
APICAL RIDGE
AMNIOTIC CAVITY
AMNION OF TAIL
TAIL
NEURAL TUBE
POSTERIOR CARDINAL VEIN
WOLFFIAN DUCT
SPLANCHNOCOEL
RECTUM
CLOACA
ALLANTOIC STALK
AMNIOTIC CAVITY
TAIL GUT
NOTOCHORD

82 — 82

Drawing of Specimen 82

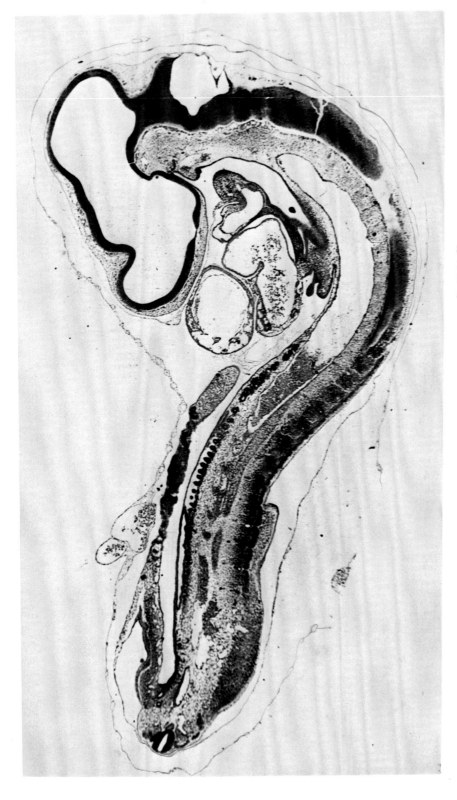

83. **Chick:**
36-somite stage,
H.L.S.
mag. 25 ×

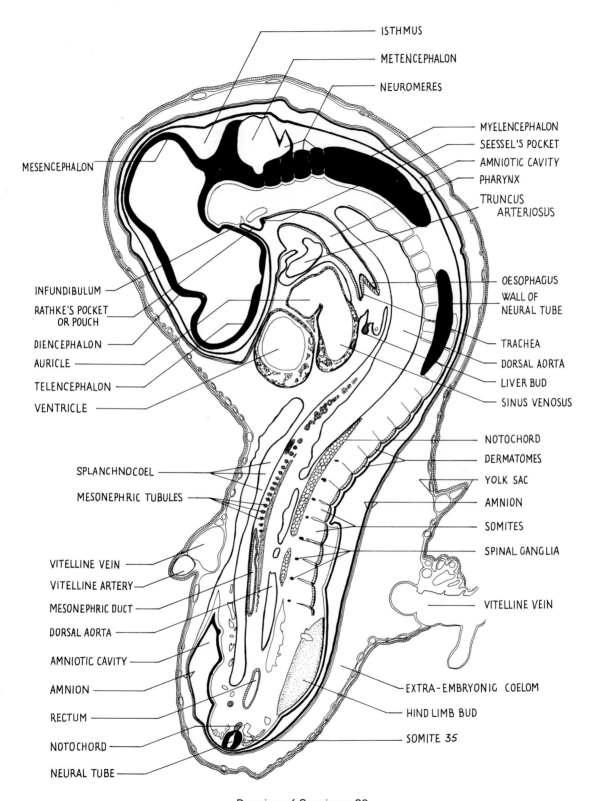

ISTHMUS

METENCEPHALON

NEUROMERES

MYELENCEPHALON

SEESSEL'S POCKET

AMNIOTIC CAVITY

PHARYNX

TRUNCUS ARTERIOSUS

MESENCEPHALON

OESOPHAGUS

WALL OF NEURAL TUBE

INFUNDIBULUM

RATHKE'S POCKET OR POUCH

DIENCEPHALON

TRACHEA

DORSAL AORTA

AURICLE

LIVER BUD

TELENCEPHALON

SINUS VENOSUS

VENTRICLE

NOTOCHORD

DERMATOMES

YOLK SAC

SPLANCHNOCOEL

AMNION

MESONEPHRIC TUBULES

SOMITES

SPINAL GANGLIA

VITELLINE VEIN

VITELLINE ARTERY

VITELLINE VEIN

MESONEPHRIC DUCT

DORSAL AORTA

AMNIOTIC CAVITY

EXTRA-EMBRYONIC COELOM

AMNION

HIND LIMB BUD

RECTUM

SOMITE 35

NOTOCHORD

NEURAL TUBE

Drawing of Specimen 83

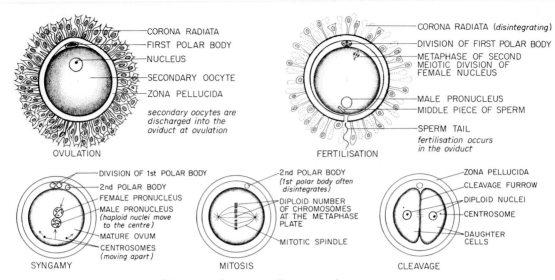

Diagrams of a mammalian secondary oocyte,
fertilised ovum and first cleavage.

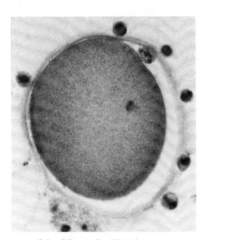

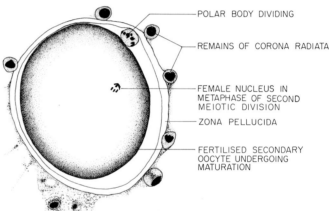

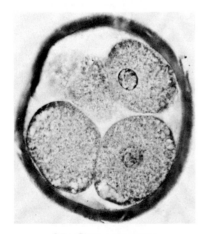

84. **Man:** fertilised ovum,
V.S. *mag. 600 ×*

Drawing of Specimen 84

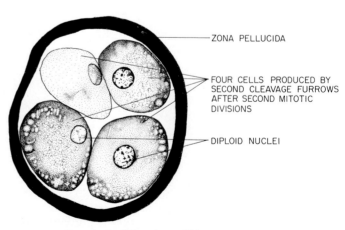

85. **Cow:** cleavage,
four-cell stage, T.S. *mag. 600 ×*

Drawing of Specimen 85

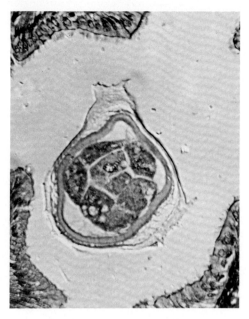

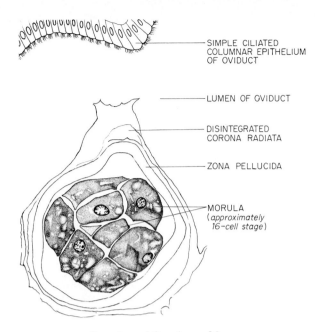

SIMPLE CILIATED
COLUMNAR EPITHELIUM
OF OVIDUCT

LUMEN OF OVIDUCT

DISINTEGRATED
CORONA RADIATA

ZONA PELLUCIDA

MORULA
(*approximately*
16-cell stage)

86. Rabbit: cleavage,
morula in oviduct, T.S. *mag. 400×*

Drawing of Specimen 86

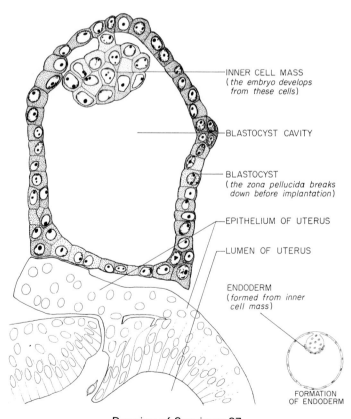

INNER CELL MASS
(*the embryo develops*
from these cells)

BLASTOCYST CAVITY

BLASTOCYST
(*the zona pellucida breaks*
down before implantation)

EPITHELIUM OF UTERUS

LUMEN OF UTERUS

ENDODERM
(*formed from inner*
cell mass)

FORMATION
OF ENDODERM

87. Guinea pig: blastocyst
in endometrium of uterus,
V.S.*mag. 460×*

Drawing of Specimen 87

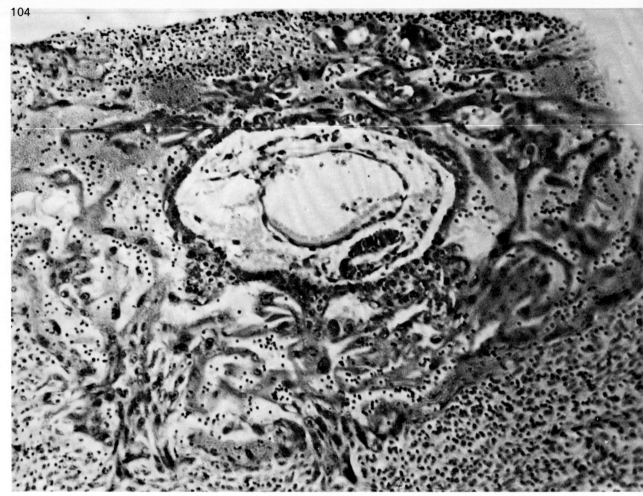

88. **Man:** previllous trophoblast implanted in endometrium of uterus,
9 to 10 days, V.S. *mag. 90 ×*

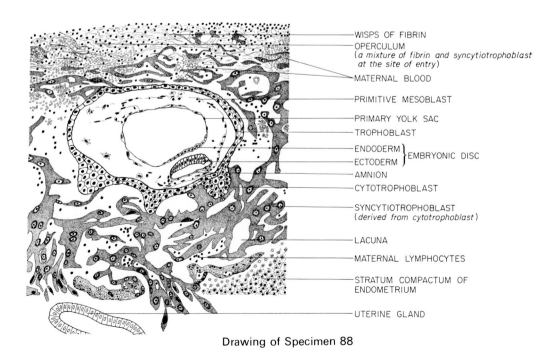

WISPS OF FIBRIN
OPERCULUM
*(a mixture of fibrin and syncytiotrophoblast
at the site of entry)*
MATERNAL BLOOD
PRIMITIVE MESOBLAST
PRIMARY YOLK SAC
TROPHOBLAST
ENDODERM
ECTODERM } EMBRYONIC DISC
AMNION
CYTOTROPHOBLAST
SYNCYTIOTROPHOBLAST
(derived from cytotrophoblast)
LACUNA
MATERNAL LYMPHOCYTES
STRATUM COMPACTUM OF
ENDOMETRIUM
UTERINE GLAND

Drawing of Specimen 88

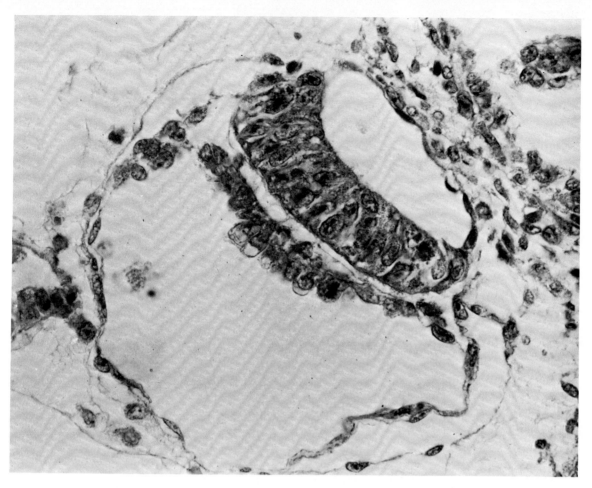

89. **Man:** formation of the amnion and yolk sac, 13 days, V.S. *mag. 480×*

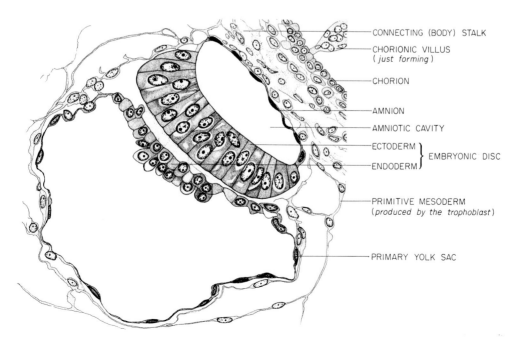

CONNECTING (BODY) STALK

CHORIONIC VILLUS
(*just forming*)

CHORION

AMNION

AMNIOTIC CAVITY

ECTODERM }
                } EMBRYONIC DISC
ENDODERM }

PRIMITIVE MESODERM
(*produced by the trophoblast*)

PRIMARY YOLK SAC

Drawing of Specimen 89

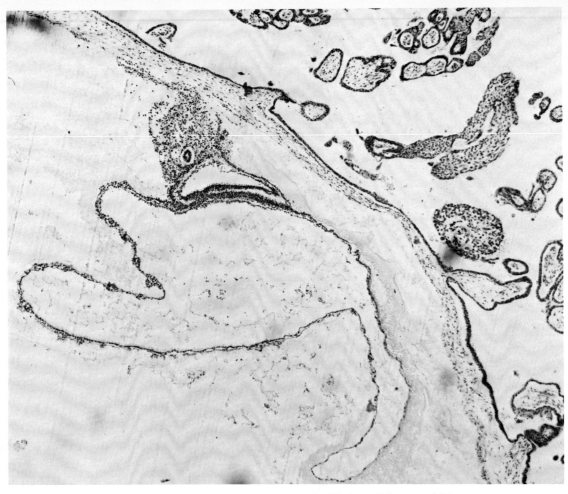

90. **Man:** primitive streak and allantois, 18 days, T.S. *mag. 360×*

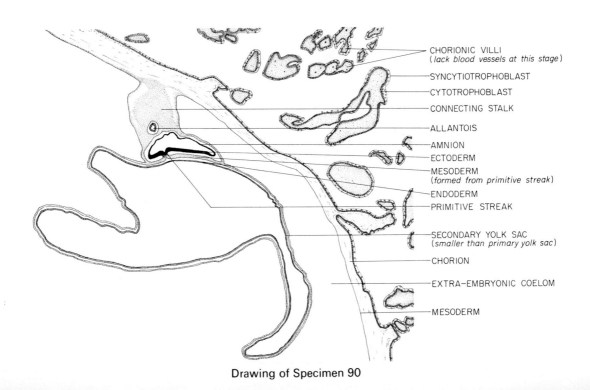

CHORIONIC VILLI
(*lack blood vessels at this stage*)

SYNCYTIOTROPHOBLAST

CYTOTROPHOBLAST

CONNECTING STALK

ALLANTOIS

AMNION

ECTODERM

MESODERM
(*formed from primitive streak*)

ENDODERM

PRIMITIVE STREAK

SECONDARY YOLK SAC
(*smaller than primary yolk sac*)

CHORION

EXTRA–EMBRYONIC COELOM

MESODERM

Drawing of Specimen 90

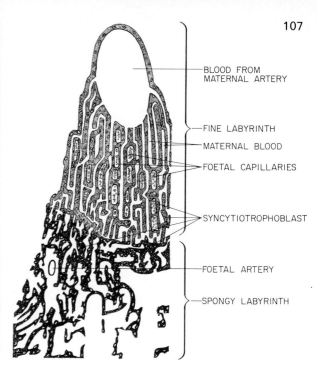

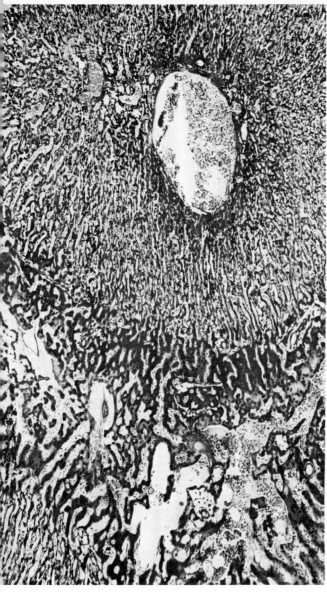

91. **Guinea pig:** placenta of the
labyrinthine type, H.S. *mag. 185×*

BLOOD FROM
MATERNAL ARTERY

FINE LABYRINTH

MATERNAL BLOOD

FOETAL CAPILLARIES

SYNCYTIOTROPHOBLAST

FOETAL ARTERY

SPONGY LABYRINTH

Drawing of Specimen 91

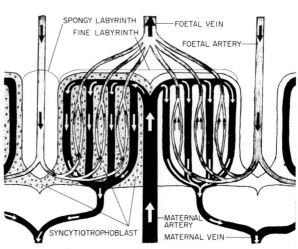

SPONGY LABYRINTH
FINE LABYRINTH
FOETAL VEIN
FOETAL ARTERY

SYNCYTIOTROPHOBLAST

MATERNAL
ARTERY
MATERNAL VEIN

Diagram showing circulation in
a labyrinthine type of placenta

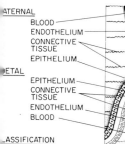

MATERNAL
BLOOD
ENDOTHELIUM
CONNECTIVE
TISSUE
EPITHELIUM

FOETAL

EPITHELIUM
CONNECTIVE
TISSUE
ENDOTHELIUM
BLOOD

CLASSIFICATION

*based on layers present;
"Ms do not support the
existence of the
hemo-endothelial type)*

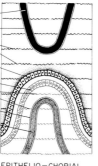

EPITHELIO—CHORIAL

*(this is the least
modified type of
placenta e.g. horse,
pig)*

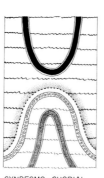

SYNDESMO—CHORIAL

*(modified by the
breakdown of the
maternal epithelium
e.g. cattle, deer)*

ENDOTHELIO—CHORIAL

*(maternal epithelium
and connective tissue
have both broken down
e.g. cat, dog)*

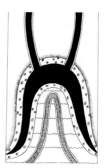

HEMO—CHORIAL

*(chorionic villi lie in
maternal blood in
intervillous spaces
e.g. man, monkey)*

HEMO—ENDOTHELIAL

*(foetal blood vessels
lie in maternal blood
in intervillous spaces
e.g. guinea pig, rat,
rabbit)*

Diagrams of different types of placenta

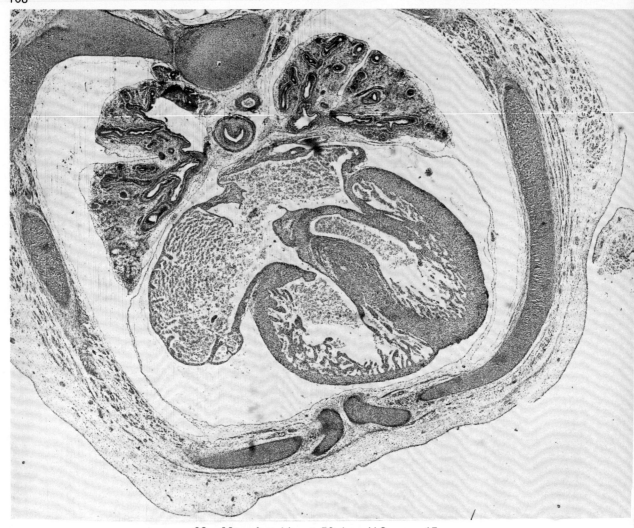

92. **Man:** foetal heart, 50 days, V.S. *mag. 15×*

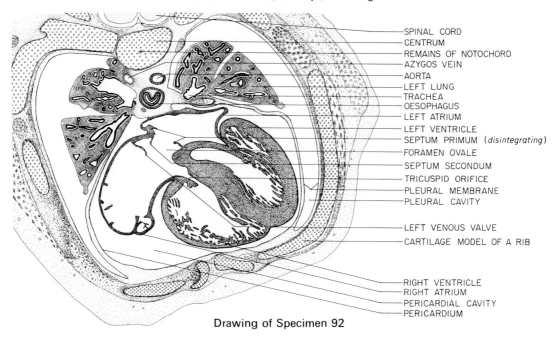

SPINAL CORD
CENTRUM
REMAINS OF NOTOCHORD
AZYGOS VEIN
AORTA
LEFT LUNG
TRACHEA
OESOPHAGUS
LEFT ATRIUM
LEFT VENTRICLE
SEPTUM PRIMUM (*disintegrating*)
FORAMEN OVALE
SEPTUM SECONDUM
TRICUSPID ORIFICE
PLEURAL MEMBRANE
PLEURAL CAVITY

LEFT VENOUS VALVE
CARTILAGE MODEL OF A RIB

RIGHT VENTRICLE
RIGHT ATRIUM
PERICARDIAL CAVITY
PERICARDIUM

Drawing of Specimen 92

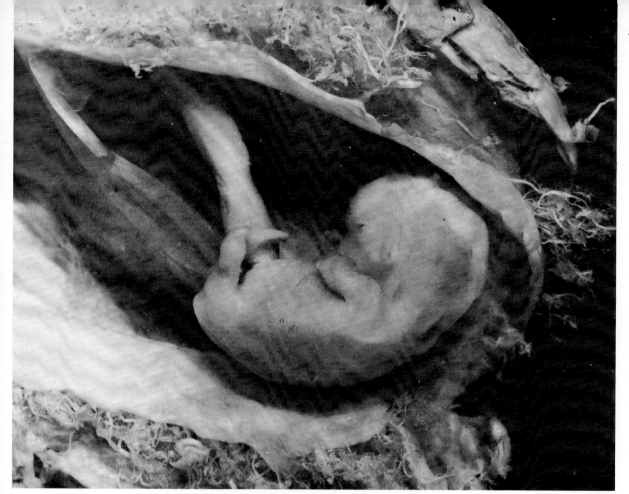

93. **Man:** 50-day foetus, W.M. *mag. 9×*

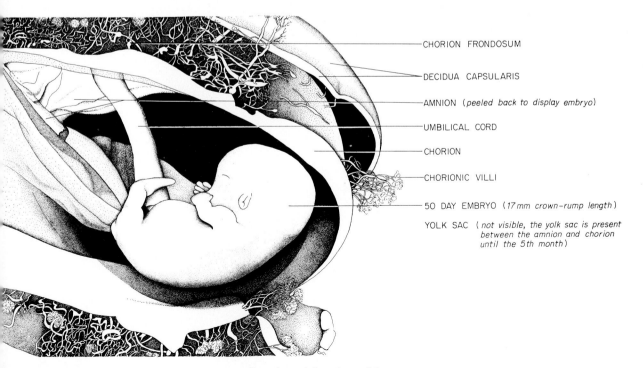

CHORION FRONDOSUM

DECIDUA CAPSULARIS

AMNION (*peeled back to display embryo*)

UMBILICAL CORD

CHORION

CHORIONIC VILLI

50 DAY EMBRYO (*17mm crown-rump length*)

YOLK SAC (*not visible, the yolk sac is present between the amnion and chorion until the 5th month*)

Drawing of Specimen 93

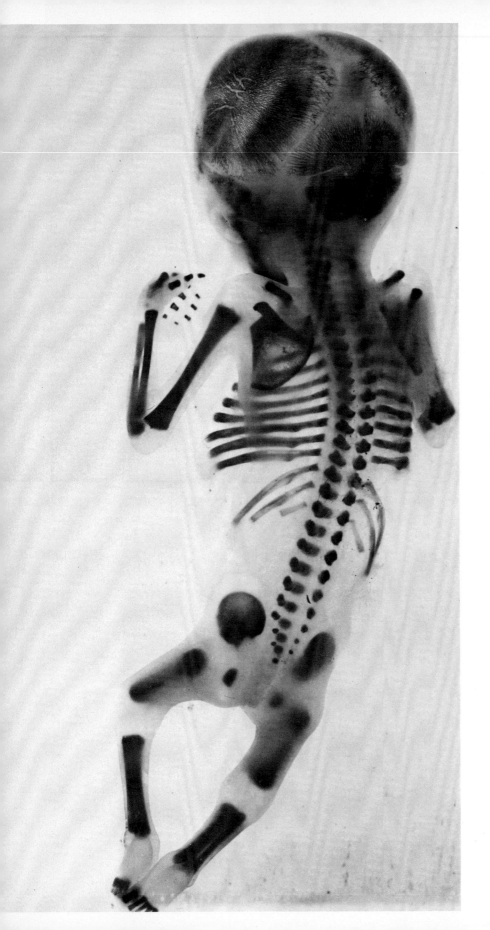

94. **Man:** foetus, alizarin preparation to demonstrate bone formation, W.M. *mag. 1.5×*

Number of somites	Stage*	Incubation time in hours according to:—				Primitive streak	Nervous system	Mesoderm, somites and kidney	Vascular system	Anterior intestinal portal
		Duval	Huettner	Patten	Lillie					
0	4	20	17–18	18	18–19	*Maximal length, 2.2 mm., i.e., 0.7 of area pellucida. Groove, pit and node present.*		*Shield shaped sheet of mesoderm spreads out laterally from the primitive streak.*		
0	5 & 6	21	19	20	19–22	*Begins to decrease in length, 1.9 mm. Notochord grows forward from node.*	*Neural plate and neural folds visible.*	*Lateral horns of mesoderm grow forward. The first somite may appear simultaneously with the formation of the head fold (stage 7).*	*Mesenchyme cells form isolated blood islands in extra-embryonic mesoderm.*	*First se to be pre ent.*
3	8–	22	23	23	25–28	*Reduced to a length of 1.5 mm.*	*Neural folds meet in brain region but do not fuse. Segmented somites joined to lateral plate mesoderm by intermediate mesoderm (nephrotome). A cavity, the myocoel, appears in somites.*	*Lateral horns grow round the mesodermless proamnion.*	*The blood islands begin to unite and the first blood corpuscles are produced within the resulting tubes.*	*Moves ba as the for gut elongates.*
5	8+	23–25	25	25–26	27–30	*1.2 mm. long.*	*Fusion of folds begins in brain region; further back neural folds meet but they splay out over the somites.*	*The cells of the somites become radially arranged about the myocoel cavities; cavity reduced by a central core of cells. Lateral horns meet anteriorly.*	*The embryo becomes linked to the blood island system by vitelline veins. Paired primordia of the heart develop together with ventral and dorsal aortae.*	*Lies poste ior to t heart prir ordia.*
10	10	29–30	30	30–31	33–38	*0.6 mm. long.*	*Except for anterior neuropore, fusion of folds is completed in the brain region. Three primary brain vesicles visible.*	*The intermediate mesoderm begins to separate off dorsally. The pronephric tubules develop from this material between somites six and ten. The first somite begins to disappear.*	*The heart primordia fuse to form a tubular heart which bends slightly to the right of the embryo. Faint and sporadic pulsation of the heart occurs.*	*May rea the fir somite.*
13	11	33–34	33	33–34	40–45	*0.4 mm. long.*	*Five brain vesicles can be seen. Anterior neuropore closes. The neural folds fuse beyond the thirteenth somite.*	*The dorso-lateral buds differentiate into pronephric tubules and the pronephric duct forms by fusion of material from the tubules. First signs of Wolffian duct.*	*The heart becomes distinctly displaced to the right. The rate and amplitude of the heart beats increase. A network of blood vessels established in area vasculosa.*	*Reaches th second somite.*
17	12+	37–41	37	38–40	46–50	*0.2 mm. long.*	*Fore brain at an angle to hind brain due to flexure. A shallow infundibulum is present.*	*Connection between somites and nephrotomes is lost. The mesonephros develops along with pronephros below the somites. Wolffian duct extends from tenth to fifteenth somite. Differentiation begins in anterior somites.*	*The heart is beating efficiently by this stage and blood circulates. The heart is S-shaped. The first aortic arch begins to develop. The dorsal aortae fuse between somites three and four. The vitelline artery can be seen between somites sixteen and seventeen.*	*Reaches t third somite.*
21	14+	43–46	43	44–48	48–52	*No longer distinguishable: contributes material to tail bud.*	*Fore brain at right angles to hind brain. Fore brain enlarges in telencephalon region.*	*Pronephros begins to disappear anterior to the eleventh somite. In the anterior somites a distinct dermatome can be seen and cells migrate from the somites and neural crests to form the sclerotomes round the notochord.*	*The atrium begins to divide into right and left auricles. The first aortic arch is established and the second begins to form. Fusion of dorsal aortae may reach somite eight. The vitelline artery is distinct between somites 17–19.*	*Reaches t fourth somite.*
24	15	44–46	48	48–50	50–55		*The telencephalon becomes distinct from the diencephalon. Rathke's pocket grows under the infundibulum.*	*The posterior somites remain undifferentiated; anteriorly somites differentiate into dermatome, myotome and sclerotome. There are eleven pairs of mesonephric tubules between somite five and sixteen.*	*Besides the two auricles heart has distinct ventricle and conus arteriosus. Two aortic arches present. Dorsal aortae fuse as far back as somite twelve. The vitelline arteries lie between somites eighteen and twenty.*	*Is in t region somites fi to six.*
27	16	48	50–52	50–55	51–56		*Telencephalon and diencephalon become separated by the velum transversum. A distinct isthmus can be seen between the mesencephalon and metencephalon.*	*Differentiation into dermatome, myotome and sclerotome reaches somite twenty. Wolffian duct and mesonephric tubules seen in trunk sections.*	*The third aortic arch appears. The dorsal aortae fuse between somites four and fourteen. The vitelline artery lies between somites 19 and 21. Vitelline veins join to form ductus venosus which opens into sinus venosus.*	*Lies b tween som ites sev and ten.*
30	17	52	58–60	55–60	52–64		*The isthmus deepens. Paired telencephalic vesicles develop. Roof of hind brain becomes very thin in myelencephalon region. Brain bent double by now.*	*Differentiation reaches the twenty-fifth somite. Wolffian duct grows back towards cloaca. Glomeruli can be seen in mesonephric tubules.*	*There are three complete aortic arches and the fourth begins to develop. The first pair of aortic arches may begin to atrophy at this stage. Dorsal aortae fused up to somite 16. Vitelline artery between somites 20 and 22.*	*Has mov back to o between somites t and twel*
36	18+	68–72	72	72	72		*The cerebral hemispheres develop from the telencephalic vesicles. The infundibulum joins with Rathke's pocket to form the pituitary.*	*Differentiation reaches the thirtieth somite. Wolffian duct reaches cloaca but may not fuse with it until later.*	*The first pair of aortic arches continue to atrophy as the fourth pair develop. Dorsal aortae fused as far back as somites 17–20. Vitelline artery is in region of somites 21–22.*	*Between somites thirteen a fourteen.*

* Hamilton & Hamburger